Introductory Notes on Planetary Science

The solar system, exoplanets and planet formation

AAS Editor in Chief

Ethan Vishniac, Johns Hopkins University, Maryland, USA

About the program:

AAS-IOP Astronomy ebooks is the official book program of the American Astronomical Society (AAS), and aims to share in depth the most fascinating areas of astronomy, astrophysics, solar physics and planetary science. The program includes publications in the following topics:

GALAXIES AND
COSMOLOGY

INTERSTELLAR
MATTER AND THE
LOCAL UNIVERSE

STARS AND
STELLAR PHYSICS

EDUCATION,
OUTREACH,
AND HERITAGE

HIGH-ENERGY
PHENOMENA AND
FUNDAMENTAL
PHYSICS

THE SUN AND
THE HELIOSPHERE

THE SOLAR SYSTEM,
EXOPLANETS,
AND ASTROBIOLOGY

LABORATORY
ASTROPHYSICS,
INSTRUMENTATION,
SOFTWARE, AND DATA

Books in the program range in level from short introductory texts on fast-moving areas, graduate and upper-level undergraduate textbooks, research monographs and practical handbooks.

For a complete list of published and forthcoming titles, please visit iopscience.org/books/aas.

About the American Astronomical Society

The American Astronomical Society (aas.org), established 1899, is the major organization of professional astronomers in North America. The membership (~7,000) also includes physicists, mathematicians, geologists, engineers and others whose research interests lie within the broad spectrum of subjects now comprising the contemporary astronomical sciences. The mission of the Society is to enhance and share humanity's scientific understanding of the universe.

Introductory Notes on Planetary Science

The solar system, exoplanets and planet formation

Colette Salyk

Vassar College, Department of Physics and Astronomy, Poughkeepsie, NY, USA

Kevin Lewis

Johns Hopkins University, Department of Earth and Planetary Sciences, Baltimore, MD, USA

IOP Publishing, Bristol, UK

ISBN   978-0-7503-2212-6 (ebook)
ISBN   978-0-7503-2210-2 (print)
ISBN   978-0-7503-2213-3 (myPrint)
ISBN   978-0-7503-2211-9 (mobi)

DOI   10.1088/2514-3433/abb198

Version: 20201201

AAS–IOP Astronomy
ISSN 2514-3433 (online)
ISSN 2515-141X (print)

British Library Cataloguing-in-Publication Data: A catalogue record for this book is available from the British Library.

Published by IOP Publishing, wholly owned by The Institute of Physics, London

IOP Publishing, Temple Circus, Temple Way, Bristol, BS1 6HG, UK

US Office: IOP Publishing, Inc., 190 North Independence Mall West, Suite 601, Philadelphia, PA 19106, USA

Supplementary python files are available at http://iopscience.iop.org/book/978-0-7503-2212-6

Contents

Preface

This text is intended to provide an introduction to the principles of planetary science to undergraduate students considering majoring in planetary science, astronomy, or a related discipline. In this text, we aim to show how the many sub-fields of planetary science are tied together with the principles of physics. We therefore assume that the reader has a knowledge of Newton's laws and mechanics, electricity and magnetism, and some introductory thermodynamics concepts (although we provide refreshers on all of these concepts as needed). The text also assumes some familiarity with vector notation.

This text provides a brief overview of several fields of planetary science, and remains at the introductory level. Therefore, by necessity, many details, and many topics, were left out. An interested reader can extend their understanding of the details by questioning the assumptions made in the text. (For example, if the text says "Assume the orbit is circular", ask yourself what might happen if the orbit were elliptical.) We also provide a list of further readings at the end of the text, that can be used to probe concepts in more depth, or to increase breadth.

Although we aimed to discuss planets and physical principles in general, we could not altogether ignore the individual personalities of the planets inside (and outside) of the solar system. We have included discussions in each chapter about the actual properties of planets, but also come nowhere near doing justice to the detailed knowledge of individual planets that has come from planetary space missions and telescopic studies. We encourage an interested reader to learn more about particular planets inside and outside of the solar system; we recommend especially that the reader visit the team webpages for various planetary missions or telescope programs. (For the record, C. Salyk's favorite planet is Saturn; K. Lewis's favorite planet is Mars.)

Notes on Figures

A majority of the figures in this text were created with python, using matplotlib (Hunter, J. D. 2007, Matplotlib: A 2D Graphics Environment, CSE, 9, 90). The scripts for these plots are available via the Publisher (and via http://iopscience.iop.org/book/978-0-7503-2212-6), and on the first author's GitHub page. Figures can therefore be customized or updated according to your needs or desires, or may be modified to demonstrate concepts not covered in this text.

Acknowledgements

The authors would like to first and foremost thank Fred Chromey, whose generously shared lecture notes provided the framework for the Vassar College course, Planetary and Space Science, and, eventually, this text. We would also like to thank all of the amazing colleagues who were willing to provide feedback on portions of our text. Our deepest thanks to Katy Garmany, Joel Green, Edwin Kite, Miki Nakajima, April Russell, Kunio Sayanagi, Terry-Ann Suer, Angelle Tanner, and Johanna Teske. We also thank an anonymous reviewer, whose careful reading of the text greatly improved our work. To be clear, any remaining errors are our fault, not theirs.

The authors would also like to thank our many planetary science and astronomy instructors and colleagues, whose courses inspired the material that appears in this book. Many thanks to Oded Aharonson, Geoffrey A. Blake, Mike Brown, Robin Canup, Andrew Ingersoll, John Johnson, John Marshall, George Rossman, Re'em Sari, and David Stevenson. Thanks also to William Hoynes and Bryan Swarthout for their assistance in getting started with publishing this work. And thank you to all of the undergraduate students at Vassar college who read early versions of this book and provided invaluable feedback on its readability.

Many excellent references made this work possible. In addition to all of the works specifically mentioned in the text, the authors wish to acknowledge the more general texts that were essential resources for our understanding of the material presented in this text: "Planetary Sciences", by Imke de Pater and Jack Lissauer, "Lecture Notes on the Formation and Early Evolution of Planetary Systems" by Philip J. Armitage, "The Formation of Planets", by Steven P. Ruden, "How Do You Find an Exoplanet?" by John Johnson, "Planetary Surface Processes" by H. Jay Melosh, and the (unpublished) lecture notes of Fred Chromey.

This text has also made use of many data sets that scientists have generously made easily accessible, even to a non-expert. We gratefully acknowledge all of those resources here, and provide acknowledgement statements requested by each of these databases below. The authors also gratefully acknowledge the creators of the open source python packages scipy, numpy, matplotlib, and astropy.

Last, but not least, the authors would like to thank their respective institutions and departments, the Department of Physics and Astronomy at Vassar College, and the Department of Earth and Planetary Sciences at Johns Hopkins University, for their support of this work.

Database-specific Acknowledgement Statements

Simulation results on Earth's atmosphere have been provided by the Community Coordinated Modeling Center at Goddard Space Flight Center through their public Runs on Request system (http://ccmc.gsfc.nasa.gov). The NRLMSISE-00 Model was developed by M. Picone, A. E. Hedin and D. Drob at the Naval Research Laboratory. This work has made use of data from the European Space Agency (ESA) mission Gaia (https://www.cosmos.esa.int/gaia), processed by the Gaia Data

Processing and Analysis Consortium (DPAC, https://www.cosmos.esa.int/web/gaia/dpac/consortium). Funding for the DPAC has been provided by national institutions, in particular the institutions participating in the Gaia Multilateral Agreement. This paper includes data collected by the TESS mission. Funding for the TESS mission is provided by the NASA Explorer Program.

This text has made use of the NASA Exoplanet Archive, which is operated by the California Institute of Technology, under contract with the National Aeronautics and Space Administration under the Exoplanet Exploration Program. Some of the data in the Archive were made available to the community through the Exoplanet Archive on behalf of the KELT project team. Some of the data in this archive are from the first public release of the WASP data (Butters et al. 2010) as provided by the WASP consortium and services at the NASA Exoplanet Archive, which is operated by the California Institute of Technology, under contract with the National Aeronautics and Space Administration under the Exoplanet Exploration Program. Some of the data in this archive are obtained by the MOA collaboration with the 1.8 metre MOA-II telescope at the University of Canterbury Mount John Observatory, Lake Tekapo, New Zealand. The MOA collaboration is supported by JSPS KAKENHI grant and the Royal Society of New Zealand Marsden Fund. These data are made available using services at the NASA Exoplanet Archive, which is operated by the California Institute of Technology, under contract with the National Aeronautics and Space Administration under the Exoplanet Exploration Program. Some of these data are from the UKIRT microlensing surveys (Shvartzvald et al. 2017) provided by the UKIRT Microlensing Team and services at the NASA Exoplanet Archive, which is operated by the California Institute of Technology, under contract with the National Aeronautics and Space Administration under the Exoplanet Exploration Program.

This research utilizes spectra acquired by Edward A. Cloutis and Carle M. Pieters with the NASA RELAB facility at Brown University; Copyright 2014, Brown University, Providence, RI; All Rights Reserved. Brown University disclaims all warranties with regard to these data, including all implied warranties of merchantability and fitness for any particular purpose. In no event shall Brown University be liable for any special, indirect or consequential damages or any damages whatsoever resulting from loss of use, data or profits, whether in an action of contract, negligence or other tortious action, arising out of or in connection with the use or performance of these data.

References

Butters, O. W., West, R. G., Anderson, D. R., et al. 2010, A&A, 520, L10
Shvartzvald, Y., Bryden, G., Gould, A., et al. 2017, AJ, 153, 61

Author biographies

Colette Salyk

Photo credit: ©Vassar College/Karl Rabe

Colette Salyk has a PhD in Planetary Science from the California Institute of Technology, and a BS in Planetary Science from the Massachusetts Institute of Technology. She is currently an Assistant Professor of Astronomy at Vassar College, where she teaches courses in planetary science, observational astronomy, and introductory physics. She studies the formation of planets using ground- and space-based telescopes.

Kevin Lewis

Photo credit: Johns Hopkins University/Kathryn Vitarelli

Kevin Lewis has a PhD in Planetary Science from the Californita institute of Technology, and BS degrees in Physics and Mathematics from Tufts University. He is currently an Assistant Professor in the Department of Earth and Planetary Sciences at Johns Hopkins University, where he teaches courses in remote sensing and the geology of the Earth and other planets. His research focuses on the geological evolution of the terrestrial planets, and in particular the climate history of Mars.

Introductory Notes on Planetary Science
The solar system, exoplanets and planet formation
Colette Salyk and Kevin Lewis

Chapter 1

Introduction

1.1 What is a Planet?

What exactly is a planet? While scientists have long pondered this question, the general public took a special interest in this question when Pluto was "demoted" in 2006, losing its planet status. How could it be that the solar system used to have nine planets and now has only eight? And what, exactly, is the newly-coined term *dwarf planet*? In our experience, planetary scientists are not too concerned about the definition of a planet, since *planet* is simply a label. Pluto is still orbiting the Sun, doing exactly what it has done for eons, whether or not it is called a planet or a dwarf planet. Nevertheless, it can be fun to discuss: what exactly is a planet?

Historically, the word planet comes from the Greek, and means "wanderer". This is because the planets of the solar system are closer to us than stars, and therefore appear to wander through the stars from an Earthling's perspective. In ancient times, observations were limited to objects that could be seen with the naked eye. Therefore, all such wanderers were also relatively large. After telescopes were developed in the 1600s, however, it became possible to find ever fainter, and therefore smaller, wanderers. And so the definition of planet was modernized to specify that the wanderers must be large to count as a planet.

In this text, we'll take an expansive, yet modern, definition of planet. For our purposes, a planet is any large object that orbits a star but is not a star itself. With this definition, we can focus our attention on the processes that affect these objects, without worrying too much about what to call them. However, even with this simple definition, it's easy to see how trying to strictly define what a planet is becomes tricky. For example, what do we mean by "large"? Is a pebble a planet? Is an asteroid a planet? Is Pluto a planet? What does it mean to orbit a star? If a moon is orbiting a star, but also orbiting a planet, does it count as a planet? What does it mean to be a star? Does a star have to be undergoing nuclear fusion of hydrogen, or does a (less massive) object that fuses the heavier deuterium atom also count as a star? What about if the object used to fuse hydrogen and has evolved so that it no longer does? Finally, it's entirely possible that something we would think of as a planet could get kicked out of its planetary system and end up flying alone through space. If this rogue planet is no longer orbiting a star, is it still a planet? Whew.

1.1.1 How Many Planets Are There?

The debate about what counts as a planet has actually been going on for a long time. In Figure 1.1, we show the accepted number of planets for the last few centuries. As of 1700, the five brightest planets were known: Mercury, Venus, Mars, Jupiter, and Saturn, and the recognition that the Earth, too, was a planet, made the total count six. Then, improvements in telescopes in the 18th century allowed for the discovery of Uranus, and then large asteroids, including Ceres, Pallas, Juno and Vesta. At that time, there was no distinction made between asteroids and planets, but as the numbers grew, they came to be recognized as their own class of objects. In the early 1800s, the definition of a planet was therefore fuzzy, with some astronomers characterizing asteroids as planets, and others characterizing them as more "minor" objects. Neptune was discovered in 1846, prompted by measured irregularities in the orbit of Uranus (Levenson 2015). Since Mercury also had an orbit that could not be explained with Newtonian Physics, the planet Vulcan was predicted to exist interior to Mercury's orbit, and promptly "discovered", in 1860. By 1915, however, it was discovered that general relativity was the real explanation for Mercury's orbit; the discovery of Vulcan had been spurious (Levenson 2015).

And then comes Pluto. Pluto was discovered at Lowell Observatory in Arizona in 1930 (Lowell Observatory 2020) and remained firmly in the planet category for the rest of the 20th century. (Perhaps due to its small size and extreme distance from the Sun, and probably its association with a cartoon dog, it became a kind of beloved underdog.) However, beginning in the early 90s (Jewitt & Luu 1993), astronomers began to detect many Pluto-like objects at similar distances from the Sun. After

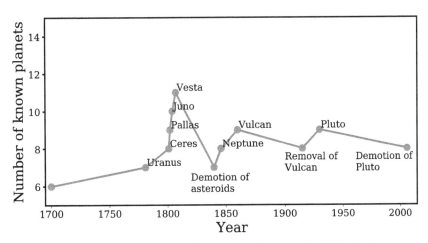

Figure 1.1. The accepted number of solar system planets versus time, since 1700. (During the discovery period of the first asteroids, scientists disagreed about what to call a planet, so these numbers should not be taken too literally.) Labels name the discovery or event that occurred in that year.

many discoveries, scientists realized that there was an entire second "belt" of objects, orbiting beyond Neptune, which has come to be called the Kuiper Belt (also known as the Edgeworth–Kuiper Belt). The definition of planet once again came into question. The issue really came to a head in 2005, when an object named Eris was discovered that appeared to be slightly larger than Pluto (Brown et al. 2005). If Pluto was large enough to be a planet, shouldn't Eris be a planet too? After many discussions between astronomers, and a vote by astronomers in the International Astronomical Union (IAU), it was decided that, officially at least, neither Pluto nor Eris would continue to be called a planet.

1.1.2 The *Official* (IAU) definition of a Planet

In order to demote Pluto, the IAU had to define what a planet was. They decided that a planet (IAU 2016):

1. *is in orbit around the Sun,*
2. *has sufficient mass for its self-gravity to overcome rigid body forces so that it assumes a hydrostatic equilibrium (nearly round) shape, and*
3. *has cleared the neighborhood around its orbit.*

They additionally defined a category of objects known as *dwarf planets* that satisfy the first two criteria, but not the third. Pluto and Eris are therefore officially characterized as dwarf planets (IAU 2016).

We shouldn't take any of this too seriously, however. Again, Pluto and Eris are still out there, living their planetary lives, whatever we choose to call them. In addition, it's pretty easy to poke holes in this definition. Does this mean that planet-like objects orbiting *other* stars are not planets? What does it mean for an object to "clear the neighborhood around its orbit"? If an object has mountains, or craters, is it still "nearly round"?

To summarize, defining what a planet is not easy, and might not be truly meaningful, though the discussion can be interesting. For the purposes of this text, we will be inclusive—welcoming all planet-like objects into the discussion.

As an interesting addendum, the most recent measurements show that Pluto is actually slightly larger than Eris (Nimmo et al. 2017)!

1.2 Solar System Overview

In this section, we'll discuss some of the bulk properties of the solar system. Ultimately, we'll try to understand how these properties reflect the physical processes of planet formation and evolution.

Here are some broad properties of the solar system we may want to understand:

- There are three broad sub-categories of planets: terrestrial planets (Mercury, Venus, Earth, Mars), gas giants (Jupiter, Saturn), and ice giants (Uranus, Neptune). Gas giants have the largest atmospheres, terrestrial planets the smallest, and ice giants something in-between. The planet sizes are shown to scale in Figure 1.2.

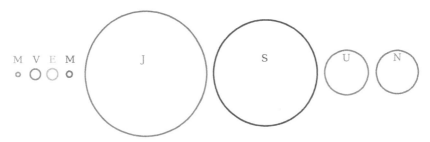

Figure 1.2. The sizes of the (IAU-defined) solar system planets, to scale, ordered from left to right according to distance from the Sun. Data from NASA (2020).

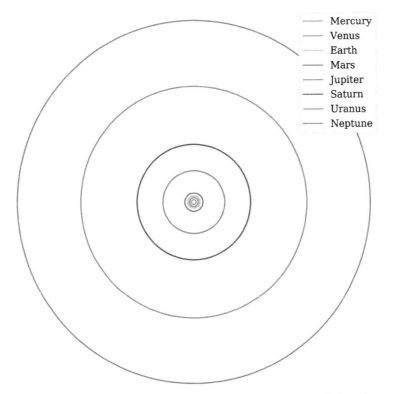

Figure 1.3. The orbital shapes for the (IAU-defined) solar system planets, to scale. The legend lists the planets from innermost to outermost orbit. Data from NASA (2020).

- The terrestrial planets orbit closest to the Sun, followed by the gas giants, and then the ice giants. The planetary orbits are shown to scale in Figure 1.3.
- There are also many smaller bodies in the solar system, including asteroids, Kuiper Belt objects, and comets. Asteroids orbit mostly between Mars and Jupiter. Kuiper Belt objects orbit primarily beyond the orbit of Neptune. Comets orbit in the outskirts of the solar system, but occasionally get kicked into the inner solar system.

- Planets orbit in ellipses with the Sun near one focus. An ellipse can be defined by the length of (half of) its long and short sides, respectively known as the *semimajor axis* (*a*) and *the semiminor axis* (*b*). These definitions are shown in Figure 1.4. A planet's closest point to the Sun is known as *perihelion* and its furthest point from the Sun is known as *aphelion*. The Earth's semimajor axis is used to define its own unit of distance, called an astronomical unit (au). In other words, Earth is defined to have a semimajor axis of 1 au. This human-centric distance is also used as a standard unit of measurement for solar system scales (for example, see the semimajor axes in Table 1.1).
- Solar system planets appear evenly spaced. A numerical representation of this observation is given by Bode's law, also called the Titius–Bode law. It turns out that the distances between the planets and the Sun are reasonably well predicted by the following expression:

$$a = 0.4 + 0.3 \times 2^n \qquad (1.1)$$

where n takes values $-\infty$, 0, 1, 2, 3, 4, 5, 6, 7, 8 and a is the object's semimajor axis in au—see Figure 1.5. It's not clear whether this "law" has any physical meaning, but we'll revisit the spacing of the planets in Chapter 9. The even spacing of the planets leads to easy-to-remember approximate distances to the giant planets. Beginning with Jupiter and ending with Neptune, the approximate semimajor axes of the giant planets are 5, 10, 20, and 30 au.
- Most of the solar system's mass is in the Sun. The mass of the Sun is $M_\odot \approx 2 \times 10^{30}$ kg. The mass of the largest planet, Jupiter, is $\approx 2 \times 10^{27}$ kg—only 1/1000th of the mass of the Sun.
- Most of the *planetary* mass in the solar system is in Jupiter. In fact, all of the other planets combined have a mass less than that of Jupiter.
- The elliptical shape of a planet's orbit is given by its eccentricity, which is defined as:

$$e = \sqrt{1 - \frac{b^2}{a^2}} \,. \qquad (1.2)$$

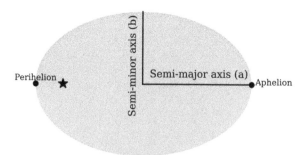

Figure 1.4. Definitions of the semimajor and semiminor axes of an ellipse. For planetary orbits, the Sun (represented by a star symbol) is near one of the two foci of the ellipse. Perihelion and aphelion represent the closest and furthest approaches of the planet to the Sun, respectively.

Table 1.1. Basic Properties of Planets (Plus Pluto and Earth's Moon) and Their Orbits, From the NASA Planetary Fact Sheet (NASA 2020)

	Mercury	Venus	Earth	Moon	Mars	Jupiter	Saturn	Uranus	Neptune	Pluto
Mass [10^{24} kg]	0.330	4.87	5.97	0.073	0.642	1,898	568	86.8	102	0.0146
Diameter [km]	4879	12,104	12,756	3475	6792	142,984	120,536	51,118	49,528	2370
Density [kg m^{-3}]	5427	5243	5514	3340	3933	1326	687	1271	1638	2095
Semimajor axis [au]	0.387	0.723	1	0.002 57	1.52	5.20	9.58	19.20	30.05	39.48
Orbital inclination (e.p.)	7.0	3.4	0.0	5.1	1.9	1.3	2.5	0.8	1.8	17.2
Orbital inclination (s.e.)	3.4	3.9	7.2		5.7	6.1	5.5	6.5	6.4	11.9
Orbital inclination (i.p.)	6.3	2.2	1.6		1.7	0.3	0.9	1.0	0.7	15.6
Orbital eccentricity	0.205	0.007	0.017	0.055	0.094	0.049	0.057	0.046	0.011	0.244
Mean temperature [K]	440	737	288	253	208	163	133	78	73	48
Rotation period [hr]	1 407.6	−5832.5†	23.9	655.7	24.6	9.9	10.7	−17.2†	16.1	−153.3†

Notes. e.p., s.e., and i.p. refer to the ecliptic plane, Sun's equatorial plane, and invariable plane, in that order.
† A negative period represents retrograde rotation.

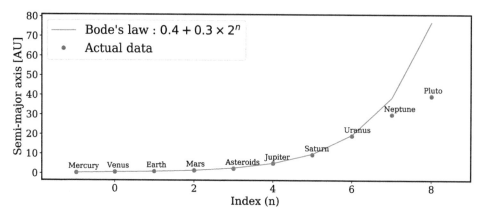

Figure 1.5. Comparison of Bode's law with actual planetary semimajor axes. No physical principles have been found to predict this "law", but the relatively even spacing of the planets may result from the planet formation process. Data from NASA (2020).

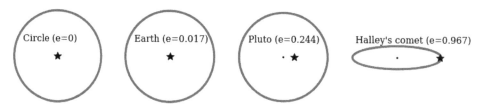

Figure 1.6. Shapes of orbits with different eccentricities. The dot marks the center of the ellipse, while the star marks the location of the Sun (one of the foci of the ellipse). Data from NASA (2020).

b is the semiminor axis of the ellipse, a is the semimajor axis of the ellipse, and e must be between 0 and 1. A circular orbit has $e = 0$ while a highly elliptical orbit would have e near 1.

The perihelion (r_p) and aphelion (r_a) distances are given by

$$r_p = a(1 - e) \tag{1.3}$$

and

$$r_a = a(1 + e). \tag{1.4}$$

The orbital eccentricities of the planets are mostly <0.1, except for Mercury and Pluto, which have $e \approx 0.2$. Comets have the highest known eccentricities in the solar system. For example, Halley's comet, which last appeared in the inner solar system in 1986, has $e = 0.967$. Figure 1.6 shows ellipses with eccentricities of 0 (a perfect circle), 0.017 (Earth), 0.244 (Pluto) and 0.967 (Halley's comet). Note that even Pluto has an orbit that looks almost circular to the eye, although the location of the Sun is noticeably shifted away from the center of the circle.

Figure 1.7. Inclinations of planetary orbits relative to the invariable plane. The unlabeled lines represent all of the other (IAU-defined) planets in the solar system. Data from NASA (2020).

- Orbital inclinations (the tilts of the orbits) are small as well. However, in order to say how small, we need to pick a reference plane from which to measure their inclination.

What are some commonly used reference frames?

The *ecliptic plane*. This plane is defined by Earth's orbit around the sun.

The Sun's equatorial plane. This plane passes through the Sun's equator.

The *invariable plane*. The plane passing through the center of mass of the solar system and perpendicular to the average *angular momentum vector* of the solar system. (An object's angular momentum vector points along its axis of rotation, following the right-hand rule).

Figure 1.7 shows the inclinations of planetary orbits relative to the invariable plane. For the most part, the solar system is very flat.

1.3 Brief Remarks on This Text

Planets come in many different shapes and sizes. Understanding their properties and interactions requires an understanding of a diverse set of sub-fields, including (but certainly not limited to) orbital and atmospheric dynamics, geology, geophysics, and chemistry. However, fundamental Physics forms the basis of all of these topics. In this text, we'll provide an introduction to the sub-fields of planetary science, using Physics as the common theme tying them together. Nevertheless, we will only be able to scratch the surface of this rich scientific field. We hope our readers will be inspired to dive deeper into their sub-field(s) of interest. But, beware—you may find yourself embarking on a lifetime of study.

1.4 Important Terms

- Planet;
- Semimajor axis;
- Semiminor axis;
- Perihelion;
- Aphelion;
- Astronomical unit (au);

- Eccentricity;
- Inclination;
- The ecliptic plane;
- The invariable plane;
- The angular momentum vector;
- Planet.

1.5 Chapter 1 Homework Questions

1. **The definition of a planet:**
   (a) Construct two lists consisting of (1) Things in the Universe that are planets, and (2) Things in the Universe that are not planets.
   (b) Given these two lists, try to come up with a definition of planets that correctly separates all objects into their appropriate list.
   (c) What did you learn while trying to come up with a comprehensive definition? Discuss any complications you considered, or challenges you encountered.
   (d) How does your definition compare to the IAU definition of a planet? Discuss the pros and cons of your definition and the IAU definition.

2. **Solar System Properties:**
   (a) Using Table 1.1; plot the mass of the planets in the solar system versus their semimajor axis (in other words, let mass be the y-axis, and semimajor axis be the x-axis). Make one plot with a linear y-axis, and one with a log y-axis (meaning that each step on the axis is ten times larger than the last). Some people say there are two types of planets in the solar system; others say there are three. Looking only at these plots, do you see evidence for distinct categories of planets? If so, how many?
   (b) Using Table 1.1, plot the density of the planets in the solar system versus their semimajor axis. Make one plot with a linear y-axis, and one with a log y-axis. What trend(s) do you see in these plots? What might be their cause?
   (c) Using Table 1.1, or additional data from your own research, make a plot of your choice. What did you learn from making this plot?

3. **Eccentricity:**
   (a) Use Equations (1.3) and (1.4) to compute the distance between the Sun and the Earth at perihelion (r_p) as well as at aphelion (r_a). By how much does these numbers differ?
   (b) Compare the difference $r_a - r_p$ to the Earth's semimajor axis. By about what *percent* does the Earth–Sun distance change over a single orbit?
   (c) Some people incorrectly believe that seasons on Earth are caused by Earth's eccentric orbit. (They are actually caused by the tilt of Earth's rotation axis.) Comment on this false hypothesis in light of the calculations you just performed.

References

Brown, M. E., Trujillo, C. A., & Rabinowitz, D. L. 2005, ApJL, 635, L97

IAU (International Astronomical Union) 2016, Resolution GA26 B5: definition of a Planet in the Solar System, http://www.iau.org/static/resolutions/Resolution_GA26-5-6.pdf

Jewitt, D., & Luu, J. 1993, Natur, 362, 730

Levenson, T. 2015, The Hunt for Vulcan (New York: Random House LLC)

Lowell Observatory 2020, The Discovery of Pluto, https://lowell.edu/the-discovery-of-Pluto/

NASA/Goddard Space Flight Center 2020, Space Science Data Coordinated Archive: Planetary Fact Sheet, https://nssdc.gsfc.nasa.gov/planetary/factsheet/

Nimmo, F., Umurhan, O., Lisse, C. M., et al. 2017, Icar, 287, 12

Introductory Notes on Planetary Science
The solar system, exoplanets and planet formation
Colette Salyk and Kevin Lewis

Chapter 2

Energy from the Sun

In this chapter, we'll discuss energy in the solar system, including: how energy is generated in the Sun via fusion, Planck's blackbody radiation curve, Wien's law, the Stefan–Boltzmann law, the definition of luminosity, the inverse square law for light, albedo, and equilibrium temperature.

The planets get their energy from several sources, including:
1. The Sun via radiation;
2. Gravitational energy of formation;
3. Decay of radioactive elements;
4. Gravitational tides (transfer of energy from orbits or spins to thermal energy).

In this chapter, we'll discuss only the energy from the Sun. In future chapters, we'll discuss the other three sources of energy.

2.1 Energy Generation in the Sun

The Sun gets its energy from nuclear fusion. Nuclear fusion is the process of fusing hydrogen (H) atoms together to produce helium (He) atoms. Figure 2.1 is a cartoon depiction of these two atoms, showing that a typical hydrogen atom has one proton in its nucleus, while a typical helium atom's nucleus has two protons and two neutrons. An atom can be described by specifying its number of protons and total mass (which is the sum of the mass of its protons and neutrons) as a subscript and superscript, respectively, in atomic mass units (where the mass of a proton is 1). So, these atoms can be written as $^1_1$H and $^4_2$He.

Given the numbers of protons and neutrons in the H and He atoms, it becomes clear that one can't simply fuse an integer number of H atoms together to make an He atom. Instead, the atoms undergo a series of fusion reactions known as the proton–proton chain. Nevertheless, the end result is essentially the conversion of 4 H atoms into one He atom. The mass of $^4_2$He is just *slightly* less than that of $4 \times {}^1_1$H.

doi:10.1088/2514-3433/abb198ch2   2-1

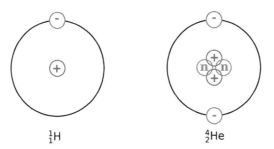

Figure 2.1. The cartoon structure of hydrogen and helium atoms. Pluses, minuses, and the letter n mark protons, electrons and neutrons, in that order. Clearly, simply adding two hydrogen atoms together won't make the ingredients of a helium atom. Instead, the fusing of hydrogen into helium involves a series of steps known as the proton–proton chain.

This difference in mass, Δm releases a small amount of energy given by Einstein's famous mass–energy equivalence equation,

$$\Delta E = \Delta mc^2. \tag{2.1}$$

An estimate of ΔE is given as a homework exercise.

Due to electrical forces that would tend to repel positively charged nuclei from each other, nuclear fusion only occurs at high pressures, in the very interior of the Sun. The energy generated in the Sun's core then gets transported to its surface via both radiation and convection. To a large extent, when discussing the effect of the Sun on planets, we can ignore all of the inner workings of the Sun, and just treat it like a simple *blackbody*—a uniform body that can be entirely characterized by its temperature and size.

2.2 Blackbody Curves and Luminosity

An ideal body that is assumed to be both a perfect absorber (i.e., it absorbs all radiation that hits it) and a perfect emitter (i.e., it emits the maximum possible energy for a body of its temperature) is known as a *blackbody*. The radiation it emits per unit time per unit frequency per unit area is known as *Planck's blackbody radiation curve*. It has the form:

$$I_\nu(\nu,\,T) = \frac{2\pi h\nu^3}{c^2}\frac{1}{\exp\frac{h\nu}{k_B T} - 1} \tag{2.2}$$

where ν is frequency, T is surface temperature, h is Planck's constant, c is the speed of light, k_B is Boltzmann's constant, and I_ν is intensity. In the SI system, I_ν has units W m^{-2} Hz^{-1}; in the cgs system, it has units erg s^{-1} cm^{-2} Hz^{-1}. Planck's blackbody radiation curve depends on *both* temperature and frequency. The curve is shown for three different temperatures in Figure 2.2. Physically speaking, this expression provides the total energy emitted by a blackbody in one second from a 1 m^2 patch of the object's surface, over a 1 Hz frequency band (if we're expressing things in SI units).

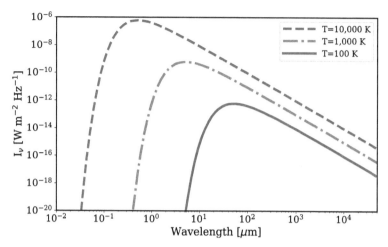

Figure 2.2. Planck blackbody radiation curves for three different temperatures. Notice how as the temperatures increase, the y values increase—a demonstration of the Stefan–Boltzmann law. In addition, the curves peak at shorter wavelengths—a demonstration of Wien's law.

Because wavelength (λ) and frequency are related by the equation $\lambda \nu = c$, one can also compute I_λ (energy per unit time per unit area per unit wavelength) and/or use ν in place of λ as the x-axis for this plot. The curves will also look different depending on whether log scales are used for the axes. Since the Planck blackbody radiation curve has so many applications in science, many of these different formats may appear in sources you read.

Note that for the three different temperatures, the curves change in overall brightness AND shift in wavelength. Also, although we cannot show this without using infinitely long paper, the curves all extend to infinity in both directions.

The dependence of the peak's location on temperature is known as *Wien's law*. In fact, the wavelength of the peak has an inverse linear relationship to temperature:

$$\lambda_{\text{peak}} \propto \frac{1}{T}, \tag{2.3}$$

or, more precisely,

$$\lambda_{\text{peak}} = \frac{2900 \ \mu\text{m K}}{T}, \tag{2.4}$$

where T is given in K. Equation (2.3) shows Wien's law as a proportion ($\propto$ means "is proportional to"), implying that $\lambda_{\text{peak}} = C\frac{1}{T}$ where C is some constant. In this text, we will use this type of reasoning (sometimes called *proportional reasoning*) often as it allows us to understand *relationships* between variables and to perform so-called "back-of-the-envelope" calculations without having to memorize constants (like 2900 μm K). Back-of-the-envelope calculations are simple calculations scientists perform to gain a qualitative, or roughly quantitative, sense of a question or system. Practice with these calculations will equip you to have discussions with scientific colleagues, to attack novel scientific questions, and to assess your more precise quantitative work.

Since I_ν provides the total energy emitted by the Sun in one second from a 1 m^2 patch of the Sun's surface, over a 1 Hz frequency band, we can integrate over all frequencies to obtain the energy emitted by the Sun in one second from a 1 m^2 patch. If we integrate I_ν over frequency (i.e., take the area under the curve in Figure 2.2), we remove the dependency on ν, but still have a function that depends on T. The derived quantity, known as *flux*, is the energy per unit time per unit area (W m^{-2} in SI units, or erg s^{-1} cm^{-2} in cgs units) emitted by a blackbody. It is given by

$$F = \int_0^\infty I_\nu \, d\nu \tag{2.5}$$

$$= \sigma T^4 \tag{2.6}$$

$$\propto T^4. \tag{2.7}$$

Equation (2.6), or the relationship given in Equation (2.7), is known as the *Stefan–Boltzmann law*, and $\sigma = 5.67 \times 10^{-8}$ W m^{-2} K^{-4} is known as the Stefan–Boltzmann constant. Qualitatively, it's important to see that the T^4 factor implies a strong dependence of flux on temperature.

If flux is the total energy emitted by a blackbody in one second from a 1 m^2 patch of the object's surface (in SI units), over all frequencies combined, then we can add up all of the patches on the body's surface to find the total energy it emits per unit time—known as its *luminosity*. In other words, just as we did to remove the dependency on ν, we could integrate the flux over the surface area of the planet to remove the dependence on area. However, since a blackbody emits uniformly from all over its surface, flux is a constant, and the integration over surface area will simply return the flux times the planet's total surface area. For a spherical object (which is a pretty good approximation for stars and planets), the surface area is $4\pi R^2$, where R is the radius of the object. Therefore, the luminosity is given by:

$$L = (\text{flux}) \times (\text{surface area}) \tag{2.8}$$

$$= \sigma T^4 4\pi R^2 \tag{2.9}$$

and the units of luminosity are J s^{-1}, or Watts (SI system), or erg s^{-1} (cgs system).

Given the surface temperature and size of the Sun, we can estimate its luminosity:

$$L_\odot = (5.67 \times 10^{-8} \text{ W m}^{-2} \text{ K}^{-4}) \, (5800 \text{ K})^4 \, (4\pi) \, (6.9 \times 10^8 \text{ m})^2 \tag{2.10}$$

$$\approx 4 \times 10^{26} \text{ W}. \tag{2.11}$$

Given this luminosity, how many fusion reactions happen per second in the Sun's core? You'll calculate this as a homework problem.

Although astronomers usually restrict themselves to talking about the luminosity of stars, this equation applies to *any* (approximately) spherical blackbody, including planets. Let's perform a back-of-the-envelope calculation to see how the luminosities of the Sun and Earth compare. The Sun's radius is about 100 times as large as the Earth's, and its temperature is about 20 times hotter. Therefore, the ratio of their luminosities is:

$$\frac{L_\odot}{L_\oplus} \approx (20)^4 (100)^2$$
$$\approx 16 \times 10^4 \times 10^4$$
$$\approx 10^9.$$

This calculation gives us a sense of why the luminosity of the Earth (and, for that matter, other planets) might reasonably be ignored when discussing the energy budget of the solar system. It's also something we'll revisit when discussing the detectability of extrasolar planets—planets orbiting stars other than the Sun.

So far, when discussing Planck's blackbody radiation curve and luminosity, we have been considering the *intrinsic* brightness of an object. This is distinct from *apparent* brightness, which also depends on how far the object is from an observer. We discuss this further in the next section.

2.3 The Inverse Square Law

We now know how much energy the Sun emits per unit time, but how much reaches the planets, and what effect does it have on each planet's temperature? The Sun is emitting in all directions, so it seems like we'd need to account for the fact that only a small fraction of that energy is intercepted by planets. In addition, we might imagine that planets closer to the Sun receive more energy than planets far from the Sun. Both of these effects are captured in the *inverse square law of light*, which says that the flux (energy per unit time per unit area) received at a distance r from a blackbody is given by:

$$F = \frac{L}{4\pi r^2} \tag{2.12}$$

where L is the luminosity of the blackbody and r is the distance to the (center of the) blackbody. The inverse square law of light comes about because the blackbody emits equally in all directions, and at a distance r the energy has spread over a sphere of surface area $4\pi r^2$. Notice that since this expression depends on the location of whoever or whatever is *receiving* the light, it is an *apparent* property of the object, rather than an *intrinsic* property.

We have now mentioned flux in two different contexts in this chapter—once when discussing the flux emitted by a body, and once when discussing the flux received at a distance r from the Sun (or any blackbody). These have the same units, but are fundamentally different quantities. The first is sometimes called "radiant flux", or "radiant exitance", while the second can be called "incident flux". Remember that the former is *intrinsic* to the body, while the latter is *apparent*.

Since we know how much energy *per unit area* reaches the planet per unit time, we can multiply by the area intercepted by the planet to obtain the total energy per unit time received by the planet. Assuming the planet is close to spherical in shape, the area intercepted by the planet is given by the cross-sectional area of the planet πR^2, where R is the radius of the planet. Why should we use the cross-sectional area (πR^2) and not half the surface area of the planet ($2\pi R^2$), since one hemisphere of the planet is exposed to the light? One way to think about this is to realize that the planet would cast a circular shadow behind it, showing that the amount of light blocked depends only on the cross section.

2.4 The Equilibrium Temperature of Planets

We now know the total energy per unit time received by the planet, but what effect does this have on the planet's temperature? To begin a calculation of a planet's temperature, we recognize that planets are essentially in *thermal equilibrium*. This means that their bulk temperatures remain constant with time and, thus, the amount of energy the planet receives must be equal to the amount of energy it emits. If it received more energy than it emitted, it would increase in temperature with time; if it emitted more energy than it received, it would decrease in temperature with time. Thus, we begin our calculation by recognizing that, in equilibrium:

Energy received by planet (per unit time) = Energy leaving planet (per unit time).

We already know that the energy per unit time received by the planet is given by $\frac{L_\odot}{4\pi r^2}\pi R^2$, but what about the energy leaving the planet per unit time? This is simply the planet's luminosity, which is given by Equation (2.9).

Therefore,

$$\text{Energy received by planet (per unit time)} = \text{Energy leaving planet (per unit time)} \tag{2.13}$$

$$\frac{L_\odot}{4\pi r^2}\pi R^2 = \sigma T_e^4 4\pi R^2. \tag{2.14}$$

Here we have called the planet's temperature T_e, to represent the temperature the planet would have in equilibrium—known as the *equilibrium temperature*. Solving for T_e, we find

$$T_e = \left(\frac{L_\odot}{16\pi\sigma r^2}\right)^{1/4}. \tag{2.15}$$

At this point, we can briefly check that the expression matches our intuition. Indeed, we would expect the temperature of the planet to increase if $L_\odot$ (the luminosity of the star) increases, and decrease if r (the distance between the star and planet) increases. We also see the interesting result that the size of the planet, given here by the radius R, doesn't matter. A bigger planet both absorbs more light AND emits more light, and the net effect is a wash.

Let's also revisit our assumptions. One key assumption we made is that the planet is a perfect blackbody—it absorbs all of the energy it receives ($\frac{L_\odot}{4\pi r^2}\pi R^2$). In general, planets do not absorb all of the light that hits their surface, but may reflect some

Table 2.1. Albedos of Solar System Planets[†].

Planet	Albedo
Mercury	0.068
Venus	0.77
Earth	0.306
Mars	0.250
Jupiter	0.343
Saturn	0.342
Uranus	0.300
Neptune	0.290

[†] Bond albedos, from the NASA Planetary Fact Sheet (NASA 2020).

instead. The fraction of light reflected is called the planet's *albedo*, A. A perfectly reflective planet would have $A = 1$, while a perfectly absorbing planet would have $A = 0$. Table 2.1 gives values of A for the solar system planets. Since A gives the fraction of light *reflected*, the total amount of energy *absorbed* by a planet is given by $\frac{L_\odot}{4\pi r^2}\pi R^2(1 - A)$.

Similarly, we assumed that the planet emits as much light as it possibly can— ($\sigma T_e^4 4\pi R^2$). In general, this is a pretty good assumption for large bodies, like planets (strictly speaking, it is a good assumption whenever the size of the body is much greater than the wavelengths of light being emitted, i.e., whenever $R \gg \lambda$). Nevertheless, to generalize our derivation of equilibrium temperature, we introduce a property known as emissivity, ϵ—the fraction of light emitted by a body compared to that of a perfect blackbody. For a perfect emitter, therefore, $\epsilon = 1$, while for something that emits no light, $\epsilon = 0$, and the light emitted by a planet is given by $\epsilon\sigma T_e^4 4\pi R^2$. Thus,

$$\frac{L_\odot}{4\pi r^2}\pi R^2(1 - A) = \epsilon\sigma T_e^4 4\pi R^2. \tag{2.16}$$

Solving for T_e, we find our final expression for equilibrium temperature:

$$T_e = \left(\frac{L_\odot(1 - A)}{16\pi\epsilon\sigma r^2}\right)^{1/4}. \tag{2.17}$$

In proportional terms, we can see that $T_e \propto r^{-1/2}$, and we have derived an important (if not too surprising) fact about the solar system: the farther a planet is from the Sun, the colder it is.

Finally, note that in this derivation, we have assumed that the planets do not have any additional sources of energy, besides the Sun, and that the planets' atmospheres do not affect their temperatures. As a homework exercise, you'll explore how well this expression approximates the true temperatures of the solar system planets, and consider whether/when these assumptions are valid. In future chapters, we'll discuss additional sources of energy, and how atmospheres can warm planetary surfaces via the greenhouse effect.

2.5 Important Terms

- Nuclear fusion;
- Blackbody;
- Planck's blackbody radiation curve;
- Wien's law;
- Proportional reasoning;
- Flux;
- Stefan–Boltzmann law;
- Luminosity;
- The inverse square law of light;
- Thermal equilibrium;
- Equilibrium temperature;
- Albedo;
- Emissivity.

2.6 Chapter 2 Homework Questions

1. Nuclear fusion:
   (a) Look up the mass of a hydrogen atom and a helium atom and use these values to estimate the energy released in a single fusion reaction, i.e., by the conversion of 4 hydrogen atoms to a single helium atom. Assume the entire Δm of the reaction is released as energy.
   (b) An energy bar you might have as a snack has about 200 Calories (= 200 kcal). Describe how the energy released in a single fusion reaction compares to the energy available in an energy bar.
   (c) The luminosity of the Sun is about 4×10^{26} W. Based on your answer to part (a), about many fusion reactions occur per second in the Sun?
   (d) Given that each reaction fuses four hydrogen atoms, approximately what mass of hydrogen is fused each second?
   (e) What mass of hydrogen has been fused in the ~4.5 billion years of the Sun's life so far?
   (f) How does your answer to part (e) compare to the mass of the entire Sun? (More precisely, what is the ratio of the mass fused so far to the total mass of the Sun?) Was this answer what you expected? (You may want to read about the expected lifetime of the Sun to provide some context.) Explain your answer.
2. Equilibrium temperature:
   (a) Plot a curve representing equilibrium temperature (in Kelvin) versus distance from the Sun (in au), assuming albedo, $A = 0$ and emissivity, $\epsilon = 1$. Use a log scale for your x-axis.

(b) Using Table 1.1, plot the mean temperatures of the planets on top of your curve.

(c) You plotted mean temperatures but of course you know (based on Earth) that temperatures can vary across the surface. This is particularly true for Mercury, which has surface temperatures ranging from 100 K to 700 K. Show these points for Mercury on your plot as well.

(d) Based on this plot, which planets have large deviations between equilibrium temperature and actual surface temperature? Discuss possible reasons for these deviations for each planet with a large deviation. (You can decide what "large" deviation means, but should discuss this choice.)

(e) Calculate the equilibrium temperature for the Earth assuming $A = 0$, $\epsilon = 1$.

(f) Some of Earth's sediments suggest that Earth underwent a nearly completely ice-covered period known as "snowball Earth". If the Earth were completely covered in ice, it may have had an albedo near 0.8. Re-compute the equilibrium temperature for the Earth, assuming $A = 0.8$. Discuss the difference between this value and your answer to the previous question, and any potential implications for life on Earth during a snowball phase.

Reference

NASA/Goddard Space Flight Center 2020, Space Science Data Coordinated Archive: Planetary Fact Sheet, https://nssdc.gsfc.nasa.gov/planetary/factsheet/

Introductory Notes on Planetary Science
The solar system, exoplanets and planet formation
Colette Salyk and Kevin Lewis

Chapter 3

Planetary Dynamics for Two Bodies

In this chapter, we will use Newton's laws to understand the motions of planets, including Kepler's laws. We'll discuss the two-body problem, as well as the energy and angular momenta of planetary orbits. With this understanding, we'll be able to understand planetary interactions that lead to shepherding. In addition, we'll understand the motion of the Sun, which will lead us to the discovery of exoplanets covered in the next chapter.

3.1 Review of Newton's Laws and Vector Notation

We'll begin this section with a review of Newton's laws, including reviewing or introducing a type of vector notation that will be especially useful for understanding planetary dynamics. First, let's review Newton's laws in a qualitative sense. They are:

1. **Newton's first law:** An object at rest stays at rest, if it experiences no net force. An object in motion stays in motion.
2. **Newton's second law:** Acceleration is proportional to force. Mass is the constant of proportionality, such that a larger mass requires a larger force to cause the same acceleration.
3. **Newton's third law:** All forces occur in action/reaction pairs. The force of object 2 on object 1 is equal in magnitude to the force of object 1 on object 2, but opposite in direction.

Recall that position is a *vector*. We can express position as the vector:

$$\vec{r} = r\,\hat{r} \tag{3.1}$$

where $\hat{r}$ is a unit vector (a vector of magnitude 1) pointing in the direction of $\vec{r}$ and r is the magnitude of the vector $\vec{r}$. If we have two vectors, $\vec{r_1}$ and $\vec{r_2}$, then $\vec{r_2} - \vec{r_1}$ points from $\vec{r_1}$ to $\vec{r_2}$ as shown in Figure 3.1. If this is not intuitive, draw yourself a diagram, like in Figure 3.1 showing the vector addition of $\vec{r_2}$ and $-\vec{r_1}$.

doi:10.1088/2514-3433/abb198ch3

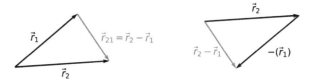

Figure 3.1. Visualizations of vectors and vector difference. Remember that you can add vectors by placing them tail to head, and drawing the vector going from the tail of the first vector to the head of the last. In addition, multiplying a vector by −1 flips it around. Finally, you can think of $\vec{r}_2 - \vec{r}_1$ as $\vec{r}_2 + (-\vec{r}_1)$.

Velocity and acceleration are also vectors:

$$\vec{v} = \frac{d\vec{r}}{dt} = \dot{\vec{r}} \tag{3.2}$$

$$\vec{a} = \frac{d\vec{v}}{dt} = \frac{d^2\vec{r}}{dt^2} = \ddot{\vec{r}}. \tag{3.3}$$

Note that we are using the dot above the vector to represent the time derivative of the vector, and two dots to represent a second derivative with respect to time.

With this notation, we can now write Newton's laws in a mathematical format.

1. **Newton's first law:** For an isolated system (one that does not experience a net external force)

$$\frac{d(m\vec{v})}{dt} = \frac{d\vec{p}}{dt} = 0, \tag{3.4}$$

where m refers to mass, and p refers to momentum. If the mass, m, is constant, then

$$\frac{d\vec{v}}{dt} = 0. \tag{3.5}$$

2. **Newton's second law:**

$$\vec{F} = m\vec{a} = m\frac{d\vec{v}}{dt} = m\frac{d^2\vec{r}}{dt^2} = m\ddot{\vec{r}}. \tag{3.6}$$

Alternatively,

$$\vec{F} = \frac{d\vec{p}}{dt}. \tag{3.7}$$

3. **Newton's third law:**

$$\vec{F}_{12} = -\vec{F}_{21} \tag{3.8}$$

where the subscript 12 means the force of 2 on 1, while the subscript 21 means the force of 1 on 2.

3.2 Two-body Interactions

Planetary systems generally consist of many objects (a central star, planets, asteroids, etc) and their collective motions cannot be described analytically (i.e., they cannot be described by a single equation—they must be computed numerically). However, the motions of two objects *can* be described analytically, and this work is known as the *two-body problem*. The goal of the *two-body problem* is to begin with Newton's laws and use them to derive an *equation of motion* describing the motion of the bodies.

The process of starting with Newton's laws and deriving an equation of motion for a system will be familiar to you if you have studied simple harmonic motion, such as the motion of a mass on a spring, or the motion of a pendulum. In the case of a spring, for example, we combine Newton's second law with Hooke's law to find the equation of motion: $-kx = m\ddot{x}$, where k is the spring constant and m is the mass of the object attached to the spring. If we solve this equation of motion, we find that the position of the mass, $x(t)$, is described by a sine or cosine curve. For the two-body problem, things are a bit more complicated, because we have two bodies moving, and the distance between the two bodies determines the magnitude and direction of the force between them. So, as we'll see in the following sections, we won't be able to derive individual equations of motion for each body, and we'll use some clever tricks to understand their collective motion.

3.2.1 Motion of the Center of Mass

When discussing the two-body problem, we consider two masses, m_1 and m_2, at positions $\vec{r}_1$ and $\vec{r}_2$, which are pulling on each other via gravitational forces only. Let's begin by considering the motion of the center of mass of the system. The center of mass is the mass-weighted center of the system. Therefore, for two masses, m_1 and m_2, the position of the center of mass is given by:

$$\vec{r}_{\text{cm}} = \frac{m_1\vec{r}_1 + m_2\vec{r}_2}{m_1 + m_2} = \frac{m_1\vec{r}_1 + m_2\vec{r}_2}{M} \tag{3.9}$$

where $M = m_1 + m_2$, and "cm" refers to the center of mass. Play around with this equation, and you can convince yourself that if the two masses are equal, the center of mass resides halfway between the two. In addition, if one mass is much larger than the other, the center of mass gets closer and closer to the more massive body.

If we take two time derivatives of this equation, we find:

$$\ddot{\vec{r}}_{\text{cm}} = \frac{m_1\ddot{\vec{r}}_1 + m_2\ddot{\vec{r}}_2}{M}. \tag{3.10}$$

Newton's second law tells us that

$$m_1\ddot{\vec{r}}_1 = \vec{F}_1 \tag{3.11}$$

and

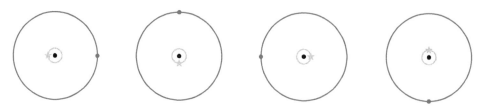

Figure 3.2. The dance of a planet (blue circle) and star (yellow star) around their mutual center of mass (black circle).

$$m_2 \ddot{\vec{r}}_2 = \vec{F}_2 \tag{3.12}$$

where $\vec{F}_1$ is the force on mass 1, and $\vec{F}_2$ is the force on mass 2.

Therefore,

$$\ddot{\vec{r}}_{cm} = \frac{\vec{F}_1 + \vec{F}_2}{M}. \tag{3.13}$$

The forces $\vec{F}_1$ and $\vec{F}_2$ are the gravitational forces between the two objects, and are an action/reaction pair. Therefore,

$$\vec{F}_1 = -\vec{F}_2 \tag{3.14}$$

and we find that:

$$\ddot{\vec{r}}_{cm} = 0. \tag{3.15}$$

We have found that the *center of mass* of the system does not accelerate. This result may sound trivial, but it has several interesting, non-trivial, implications. It implies that the two bodies (for example, a star and a planet) undergo a very particular kind of dance, in which if one of the bodies moves, the other one must move in such a way that the center of mass remains stationary.[1] For example, as shown in Figure 3.2, if a planet is undergoing orbital motion around a star, the star cannot be stationary, but must also orbit, such that the center of mass remains fixed. This fact will become crucially important when we learn about the detection of extrasolar planets. Finally, we can see that, in order to keep the center of mass in the same location, the star and planet must orbit at the same angular rate, and, therefore, have the same orbital period.

3.2.2 Equation of Motion

To derive an equation of motion for this system, we first need to recall *Newton's law of gravitation*, which we'll express in a vector form. The force on mass 1 by mass 2 is:

$$\vec{F}_1 = \frac{G m_1 m_2}{r_{21}^2} \hat{r}_{21} \tag{3.16}$$

---

[1] Strictly speaking, following Newton's first law, the center of mass can also move with constant velocity.

where $\vec{r}_{21} = \vec{r}_2 - \vec{r}_1$ and points from mass 1 to mass 2, as shown in Figure 3.1. The force on mass 2 by mass 1 is:

$$\vec{F}_2 = -\frac{Gm_1m_2}{r_{21}^2}\hat{r}_{21}. \tag{3.17}$$

To derive the equation of motion, we begin with a seemingly arbitrary computation:

$$m_1m_2\ddot{\vec{r}}_2 - m_1m_2\ddot{\vec{r}}_1. \tag{3.18}$$

Given Newton's second law, we can say that this is equivalent to:

$$m_1m_2\ddot{\vec{r}}_2 - m_1m_2\ddot{\vec{r}}_1 = m_1\vec{F}_2 - m_2\vec{F}_1. \tag{3.19}$$

Since $\vec{F}_1 = -\vec{F}_2$,

$$m_1m_2\ddot{\vec{r}}_2 - m_1m_2\ddot{\vec{r}}_1 = m_1\vec{F}_2 + m_2\vec{F}_2 \tag{3.20}$$

$$= (m_1 + m_2)\vec{F}_2 \tag{3.21}$$

$$m_1m_2(\ddot{\vec{r}}_2 - \ddot{\vec{r}}_1) = (m_1 + m_2)\left(-\frac{Gm_1m_2}{r_{21}^2}\hat{r}_{21}\right). \tag{3.22}$$

Dividing both sides by $m_1 + m_2$, and remembering that $\vec{r}_{21} = \vec{r}_2 - \vec{r}_1$, we find

$$\frac{m_1m_2}{m_1 + m_2}\ddot{\vec{r}}_{21} = -\frac{Gm_1m_2}{r_{21}^2(m_1 + m_2)}(m_1 + m_2)\hat{r}_{21}. \tag{3.23}$$

Here, we'll define a new variable, μ, known as the *reduced mass*

$$\mu = \frac{m_1m_2}{m_1 + m_2} \tag{3.24}$$

and again let $M = m_1 + m_2$. Our final equation of motion then becomes:

$$\mu\ddot{\vec{r}}_{21} = -\frac{G\mu M}{r_{21}^2}\hat{r}_{21}. \tag{3.25}$$

Let's review what we've done here. We wanted to determine the motion of two masses interacting with each other via gravity alone. However, we could not independently derive an equation of motion for each mass, since the motion of one body is affected by the location of the other, and vice versa. Using a few algebraic tricks, and Newton's laws, we instead derived a single equation of motion for the *relative* position of the two masses, which we called $\vec{r}_{21}$. Therefore, although we do not independently know how each object moves through space, we can describe how the two objects dance around each other. Note that if we look carefully at Equation (3.25), we see that it actually takes the form of Newton's second law, but as if there were an object with mass μ experiencing a gravitational force of magnitude $\frac{G\mu M}{r_{21}^2}$.

The solutions to this equation of motion, $\vec{r}_{21}(t)$, include parabolas, hyperbolas, circles, or ellipses, with the exact solution depending on the initial conditions. These shapes are known as *conic sections*, since they can all be formed by taking slices through a cone. Circles and ellipses are closed orbits, while parabolas and hyperbolas represent open orbits, where one object would fly by another, never to return again. It might help to think of a simple roller coaster as an analogy to understand how initial conditions affect the final motion. If one releases a roller coaster car near the bottom of a dip, the car will simply oscillate around the dip (analogous to a closed planetary orbit). Alternatively, if one releases a roller coaster car high above a dip, it may make it out of the dip on the other side, and never return (analogous to an open orbit). Coming back to planets, if energies are low, the objects are bound into orbit, and the solution to the equation of motion takes the shape of a circle or ellipse. In the next sections, we'll discuss these closed orbits in more detail.

3.2.3 Consequences for Unequal Mass Systems

For the case of a planet orbiting a star, we can assume that $m_2 \ll m_1$ and the equation of motion reduces in some interesting ways. First, let's rewrite Equation (3.25) to find the motion of mass 1 about the center of mass. We'll call this vector, the position of mass 1 relative to the center of mass, $\vec{r}'_1$, and we'll call the position of the center of mass $\vec{r}_{cm}$.

$$\vec{r}'_1 \equiv \vec{r}_1 - \vec{r}_{cm} \tag{3.26}$$

$$= \vec{r}_1 - \frac{m_1 \vec{r}_1 + m_2 \vec{r}_2}{m_1 + m_2} \tag{3.27}$$

$$= \frac{(m_1 + m_2)\vec{r}_1 - m_1 \vec{r}_1 - m_2 \vec{r}_2}{m_1 + m_2} \tag{3.28}$$

$$= \frac{m_2(\vec{r}_1 - \vec{r}_2)}{m_1 + m_2} \tag{3.29}$$

$$= -\frac{\mu}{m_1} \vec{r}_{21}. \tag{3.30}$$

This result shows us that the motion of mass 1 about the center of mass has the opposite direction as $\vec{r}_{21}$, and its magnitude is modified by a factor $\frac{\mu}{m_1}$. We can do a similar calculation for the motion of mass 2 about the center of mass.

$$\vec{r}'_2 \equiv \vec{r}_2 - \vec{r}_{cm} \tag{3.31}$$

3-6

$$= \vec{r}_2 - \frac{m_1 \vec{r}_1 + m_2 \vec{r}_2}{m_1 + m_2} \tag{3.32}$$

$$= \frac{(m_1 + m_2)\vec{r}_2 - m_1 \vec{r}_1 - m_2 \vec{r}_2}{m_1 + m_2} \tag{3.33}$$

$$= \frac{m_1(\vec{r}_2 - \vec{r}_1)}{m_1 + m_2} \tag{3.34}$$

$$= \frac{\mu}{m_2} \vec{r}_{21}. \tag{3.35}$$

Here, we find that the motion of mass 2 about the center of mass has the same direction as $\vec{r}_{21}$, but its magnitude is modified by a factor $\frac{\mu}{m_2}$.

For the case of a planet orbiting a star, $m_2 \ll m_1$. What do these factors imply in that case? If $m_2 \ll m_1$, then

$$\mu = \frac{m_1 m_2}{m_1 + m_2} \tag{3.36}$$

$$\mu \approx \frac{m_1 m_2}{m_1} \tag{3.37}$$

$$\approx m_2. \tag{3.38}$$

Then, Equation (3.38) becomes

$$\vec{r'}_2 \approx \frac{m_2}{m_2} \vec{r}_{21} \tag{3.39}$$

$$\approx \vec{r}_{21} \tag{3.40}$$

while Equation (3.30) becomes

$$\vec{r'}_1 \approx -\frac{m_2}{m_1} \vec{r}_{21}. \tag{3.41}$$

Since $\vec{r'}_2 \approx \vec{r}_{21}$, we see that the motion of the small mass, mass 2, is almost the same as whatever solution we found to the equation of motion for the two-body problem. Thus, mass 2 undergoes motion in the shape of a hyperbola, parabola, ellipse, or circle, depending on the initial conditions. Although mass 2 is actually orbiting the center of mass of the system, Equation (3.40) tells us that it's almost as if mass 2 is simply orbiting mass 1.

On the other hand, mass 1 always has a position opposite that of mass 2, and undergoes motion diminished in magnitude by a factor $\frac{m_2}{m_1}$. If we recall the "dance" that objects do about the center of mass, as shown in Figure 3.2, we now know that the orbit of mass 1 about the center of mass is a factor of $\frac{m_2}{m_1}$ smaller in size. As an example, for Jupiter and the Sun, this factor is about 1000, so the orbit of the Sun around the center of mass has a radius 1000× smaller than that of Jupiter about the center of mass.

3.3 Kepler's Laws of Planetary Motion

In this section, we will provide some simplified discussions of *Kepler's laws of planetary motion*, including some non-rigorous derivations that provide physical intuition about their origins. Although we'll show how Kepler's laws can be derived from Newton's laws of motion and gravity, it's interesting to realize that the statement of Kepler's laws preceded the work by Newton—in fact, Kepler's death preceded Newton's birth. Kepler's laws were based purely on empirical observations of planetary positions made by Johannes Kepler's boss, Tycho Brahe.

Kepler's laws are:
1. A planet's orbit is an ellipse with the Sun located at one focus of the ellipse.
2. The line connecting the planet and the Sun sweeps out equal areas in equal intervals of time.
3. The planet's orbital period, P, and its semimajor axis, a, are related. If the period is given in units of years and the semimajor axis in units of au:

$$P_{\text{yr}}^2 = a_{\text{au}}^3. \tag{3.42}$$

3.3.1 Kepler's First Law

Kepler's first law follows from our discussion in Sections 3.2.2 and 3.2.3. The equation of motion shows us that bound orbits are ellipses (or circles—a special case of an ellipse with $e = 0$). When the masses are very unequal ($m_2 \ll m_1$), we find that the larger mass moves very little, and is close to the center of mass of the system. On the other hand, the smaller mass has a much larger orbit, appearing to orbit around the larger mass. Notice, therefore, that Kepler's first law is only an approximation, since the Sun's position is not truly fixed at one focus of the ellipse. However, it's a pretty good approximation, given the large mass difference between the Sun and planets.

3.3.2 Kepler's Second Law

Kepler's second law follows from the *conservation of angular momentum*, which says that angular momentum is conserved in the absence of any external torques. The angular momentum of a planet at a position $\vec{r}$ with momentum $\vec{p}$ is given by $\vec{L} = \vec{r} \times \vec{p}$. If the mass of the planet, m_p, does not change, this can be rewritten as $\vec{L} = m_p \vec{r} \times \vec{v}$. The velocity vector, $\vec{v}$, can be broken into two components, $\vec{v} = v_r \hat{r} + v_\theta \hat{\theta}$, where $v_r \hat{r}$ points along the orbital radius, while $v_\theta \hat{\theta}$ points perpendicular to that, as shown in Figure 3.3. The cross product above implies that only the component of the velocity perpendicular to the radius vector contributes to angular momentum. Therefore, the magnitude of the angular momentum vector can be

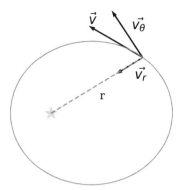

Figure 3.3. A planet's velocity vector, $\vec{v}$, can be broken into a component along its orbital radius $v_r\hat{r}$, and another component perpendicular to that, $v_\theta\hat{\theta}$.

Figure 3.4. A small arc of length $r\Delta\theta$ is swept out in a time interval Δt. Along with lines connecting the planet and star, this forms a triangle of base $r\Delta\theta$ and height r, where r is the distance between the planet and star. For small enough time intervals, we can ignore the fact that r changes with time.

rewritten as $L = m_p r v_\theta$, and, due to conservation of angular momentum, $m_p r v_\theta =$ constant. If m_p does not change, then it is also true that $r v_\theta =$ constant.

This already provides us some intuition about Kepler's second law. When the planet is close to the Sun, r is smaller so v_θ must be larger, while when the planet is far from the Sun, r is larger so v_θ must be smaller. In fact, the planet moves fastest at perihelion (closest approach to the Sun), and slowest at aphelion (furthest position from the Sun).

Now consider the motion of the planet in its orbit, letting it sweep out a small arc, as shown in Figure 3.4. In a very small interval of time, Δt, the arc is straight, has length $r\Delta\theta$, and forms a small triangle with area

$$\Delta A = \frac{1}{2}(r)(r\Delta\theta). \tag{3.43}$$

For this very small interval of time, $v_\theta = r\frac{\Delta\theta}{\Delta t}$. Therefore, this can be rewritten as

$$\Delta A = \frac{1}{2}r v_\theta \Delta t. \tag{3.44}$$

Rearranging this equation, we find that

$$rv_\theta = 2\frac{\Delta A}{\Delta t}. \tag{3.45}$$

We know that, due to conservation of angular momentum, $rv_\theta = $ constant. Therefore,

$$\frac{\Delta A}{\Delta t} = \text{constant}. \tag{3.46}$$

In words, in a time Δt, the orbit always sweeps up the same amount of area, ΔA. This is demonstrated in Figure 3.5.

Have a closer look at Figure 3.5 to consider its implications. Clearly, the planet has to move a farther distance on the blue path than on the red path. However, since these arcs have the same area, they are swept out in the same amount of time. Therefore, the planet must move faster while making the blue path as compared to while making the red path. This is the essence of Kepler's second law.

3.3.3 Kepler's Third Law

For Kepler's third law, which is a relationship between a planet's semimajor axis, a, and its orbital period, P, we'll do a simplified derivation for circular orbits. Kepler's third law of planetary motion follows from Newton's second law and Newton's law of gravitation. For circular motion, Newton's second law states that

$$F = m\frac{v^2}{r}, \tag{3.47}$$

where m is the mass of the planet in orbit, and v and r are the speed and radius of the orbit, respectively.

Newton's law of gravitation says that

$$F = G\frac{M_\odot m}{r^2}, \tag{3.48}$$

where $M_\odot$ is the mass of the Sun, and G is the gravitational constant.

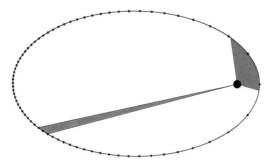

Figure 3.5. Visualization of Kepler's second law. Dots mark the path taken by an orbiting body, with equal time intervals (Δt) between each dot. The red and blue triangles are swept out in equal intervals of time, and have equal areas. Therefore, the body moves faster when it is closer to the star than when it's far from the star.

Combining these two laws, and using the fact that orbital speed is given by $v = \frac{2\pi r}{P}$, we find that:

$$m\frac{v^2}{r} = G\frac{M_\odot m}{r^2}$$

$$\left(\frac{2\pi r}{P}\right)^2 \frac{1}{r} = G\frac{M_\odot}{r^2}$$

$$\frac{GM_\odot P^2}{4\pi^2} = r^3.$$

Recognizing that $a = r$ for a circular orbit, we find:

$$\frac{GM_\odot P^2}{4\pi^2} = a^3 \qquad (3.49)$$

where a is the semimajor axis of the planet's orbit, P is its period, and $M_\odot$ is the mass of the Sun.

Kepler's third law is sometimes further simplified as

$$P^2 \propto a^3 \qquad (3.50)$$

or

$$P_{\mathrm{yr}}^2 = a_{\mathrm{au}}^3, \qquad (3.51)$$

where P_{yr} is the period of the orbit, in years, and a_{au} is the semimajor axis of the orbit, in au.

Although this simplified "proof" assumes a circular orbit, the result actually holds true regardless of the eccentricity of the planet's orbit. Therefore, the period of the planetary orbit depends only on its semimajor axis, and not on its shape. In addition, this law implies that as we get farther from the Sun, the period of a planetary orbit increases. As a homework problem, you will also find that Kepler's third law implies that planets farther from the Sun move more slowly (have lower speeds) than planets close to the Sun. Specifically, a planet's average speed,

$$v = \sqrt{\frac{GM_\odot}{a}}. \qquad (3.52)$$

3.4 Orbital Energy

In this section, we'll discuss the energy associated with an orbit. An understanding of orbital energy will help us predict some unusual properties of orbital motion (discussed in this chapter's Homework Question 3), and will eventually lead to an understanding of shepherding satellites and the analogous opening of gaps in protoplanetary disks (discussed in Chapter 9).

To begin this discussion, let's review the concept of potential energy. As you learn in introductory physics, every conservative force has an associated potential energy. For example, a mass on a spring moving along the x-axis experiences a force $F_x = -kx$ and there is an associated potential energy $U_{sp} = \frac{1}{2}kx^2$. Near the surface of the Earth (where the distance above the surface of the Earth is much smaller than the radius of the Earth, i.e., $y \ll R_\oplus$), the force of gravity is approximately $F_y = -mg$, and this has an associated potential energy $U_g = mgy$. In both cases, it is straightforward to see that the force is the negative of the derivative of the potential energy with respect to position, For the spring, $F_x = -\frac{dU}{dx}$ and for gravity, $F_y = -\frac{dU}{dy}$.

Using this fact, we can obtain a more general form of the potential energy associated with gravity, by examining Newton's law of gravitation. Given a force of gravity between two masses, m_1 and m_2 separated by a distance r,

$$F_r = -\frac{Gm_1m_2}{r^2} \tag{3.53}$$

we find an associated *gravitational potential energy*

$$U_G = -\frac{Gm_1m_2}{r}. \tag{3.54}$$

Here we use the subscript "G" to distinguish it from U_g—the latter being the potential energy associated with gravity only near the surface of the Earth. We expect and can confirm that $F_r = -\frac{dU}{dr}$:

$$-\frac{dU}{dr} = -(-Gm_1m_2)(-1)(r^{-2})$$

$$= -\frac{Gm_1m_2}{r^2}.$$

The total *orbital energy* for a planet is the sum of kinetic and potential energy,

$$E_{\text{orb}} = \frac{1}{2}m_p v^2 - \frac{GM_\odot m_p}{a} \tag{3.55}$$

where we have set $r = a$ for the planetary orbit and assume the two masses are a planet and the Sun. Substituting for v using Equation (3.52), we find:

$$E_{\text{orb}} = \frac{1}{2}m_p \frac{GM_\odot}{a} - \frac{Gm_p M_\odot}{a} \tag{3.56}$$

$$= -\frac{GM_\odot m_p}{2a}. \tag{3.57}$$

It's helpful to visualize this energy as a function of a, as in Figure 3.6. Note that as $a \to \infty$, $E_{\text{orb}} \to 0$, while as $a \to 0$, $E_{\text{orb}} \to -\infty$.

We can also see that the energy is always negative. Remember that energy is always a relative quantity. For example, if we ask what the gravitational potential

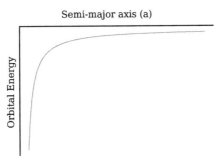

Figure 3.6. Orbital energy as a function of semimajor axis (in arbitrary units). All energies are less than zero, which is a convention for bound orbits. However, notice that orbital energies increase with increasing semimajor axis.

energy is of someone standing on top of a hill, we can define $y = 0$ to be the top of the hill or the bottom of the hill, and we'll get two different values for potential energy; nevertheless, it'll always be the case that the person has more potential energy standing on top of the hill than standing at the bottom of the hill.

For orbital energy, the energy is chosen such that it has a value of zero at $a = \infty$, to represent the border between the orbit being bound and unbound, and making the energy of all bound orbits negative. The energy becomes *less negative*, and therefore, larger, as a increases. Thus, planets far from the Sun have more total orbital energy than planets close to the Sun. As you'll explore for homework, this exposes a strange fact about orbits. Planets in larger orbits have more total energy, but also move more slowly.[2]

3.5 Important Terms

- Newton's laws;
- the two-body problem;
- equation of motion;
- center of mass;
- Newton's law of gravitation;
- reduced mass;
- conic sections;
- Kepler's laws of planetary motion;
- conservation of angular momentum;
- gravitational potential energy;
- orbital energy.

---

[2] The sign of the energy can cause some confusion that should be avoided. If you look at the absolute value of the energy, it decreases with increasing semimajor axis. However, the absolute value of energy is not a physically meaningful quantity, so we should not use it to get any intuition about orbital energy.

3.6 Chapter 3 Homework Questions

1. **Center of mass of the solar system:**
   (a) Calculate the maximum distance of the center of mass of the solar system from the center of the Sun. Use a theoretical (but unlikely!) point in time when all planets are aligned in their orbits. Place the center of the Sun at the origin, and treat the Sun and planets as point masses. Let the distance to each planet equal its semimajor axis. Compare your answer to some appropriate physical size scale (like the radius of the Sun, or the radius of Jupiter, or the radius of the Earth).
   (b) Calculate the distance from the Sun of the center of mass of a system consisting only of the Sun and Jupiter. Compare this answer to part (a) and compare the *difference* of your answers to parts (a) and (b) to some appropriate physical size scale. Comment on the appropriateness of approximations of the Sun's motion that only consider the gravitational effects of Jupiter.
   (c) Assuming that Jupiter is the only planet, and that the Sun is always the distance you derived in part (b) from the center of mass (i.e., assume it has a circular orbit around the center of mass of the Sun–Jupiter system), estimate the orbital speed of the Sun. (Hint: Think carefully about the Sun's orbital period.)

2. **Kepler's third law:**
   (a) $P^2 \propto a^3$ means that $P^2 = Ca^3$, where C is a constant. Compute the value of this constant in SI units. Be sure to also state the units of the answer.
   (b) For P in years and a in au, compute the value of the constant C. Be sure to state your answer with proper units.
   (c) Use Kepler's third law to derive the expression for orbital speed in Equation (3.52), assuming a circular orbit. What is the speed of the Earth's orbit around the Sun?
   (d) To what extent would Kepler's third law be the same, or different, in a planetary system in which the central star has twice the mass of the Sun?

3. **Weirdness of planetary orbits:** In this problem, we will investigate some weird consequences of orbital dynamics.
   (a) Think back to your introductory physics class and remind yourself what the definition of *Work* is, and what it means to do positive or negative work. In your own words, describe Work, the sign conventions, and how work relates to energy.
   (b) Remind yourself why the definition of work includes a dot product and describe this in your own words. Is work done on an object if you give it a push in a direction perpendicular to its motion?

   Use your understanding of work and the equation for orbital energy, $E = -GM_\star m_p/(2a)$ for the next two questions.

(c) Suppose you strap a rocket to the Earth and apply a force of 10^{20} N over a distance of 10^{13} m that magically always acts in a direction perpendicular to Earth's motion. How much work is done on the Earth? After the rocket is done firing, what is the new semimajor axis of the Earth?

(d) Suppose you strap a rocket to the Earth and apply a force of 10^{20} N over a distance of 10^{13} m that is magically always in the same direction as Earth's motion. How much work is done on the Earth? After the rocket is done firing, what is the new semimajor axis of Earth's orbit in astronomical units (au)? Is the new semimajor axis larger or smaller than the old semimajor axis?

(e) Given the new semimajor axis and Equation (3.52), does Earth now have a faster or slower average speed, as compared to before it got pushed? (No need to calculate the speed, just discuss whether it's faster or slower than before.) Discuss how this compares to your everyday experience with how giving something a push in the direction of its motion would affect its speed.

Introductory Notes on Planetary Science
The solar system, exoplanets and planet fomation
Colette Salyk and Kevin Lewis

Chapter 4

More Complicated Dynamics: More Than Two Bodies, and Non-point Masses

In the previous chapter, we discussed the motion of two bodies, and considered the limit where one of the bodies is much less massive than the other. In this chapter, we'll consider systems of more than two bodies. In addition, we previously treated all objects as point masses. In this chapter, we'll consider what happens when we take into account an object's finite size.

4.1 Lagrange Points

4.1.1 Description of Lagrange Points

We'll first begin by considering a system of three bodies interacting only via gravity, for which $m_1 \gg m_2 \gg m_3$, and in which m_2 travels around m_1 in a circular orbit. This is known as the *circular restricted three body problem*. Systems that are well-approximated by the circular restricted three body problem would include, for example, the Sun, Jupiter and Europa, or the Sun, Earth, and Earth's Moon, or the Earth, Earth's Moon, and a human-made satellite.

To begin this problem, we imagine that we are sitting in a frame of reference that is rotating at the same rate as m_2 is orbiting m_1, such that both m_1 and m_2 appear stationary. Importantly, this frame is accelerating—it is not an inertial frame—and so objects inside of the frame experience so-called *fictitious forces*. One of these forces is quite intuitive—it is the feeling we get on a carousel that we are being pushed toward the outer edge of the circle. This is often referred to as the centrifugal force, but, rather than being a true force, it is a consequence of an object trying to travel in a straight line, while its environment is rotating beneath it.

The second fictitious force experienced in a rotating frame is somewhat less intuitive, but can be seen clearly with the experiment shown in Figure 4.1. In this figure, we imagine a carousel rotating counterclockwise, as shown by the curved arrow, and people sitting at locations 1–4 *on the carousel*. A ball is released by the

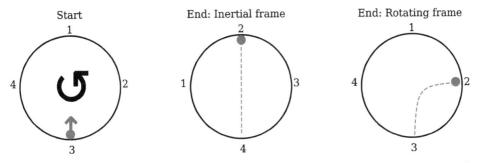

Figure 4.1. Visualization of the Coriolis effect (National Geographic 2014). A ball is thrown on a rotating carousel from position 3 upwards (left panel). In the amount of time it takes for the ball to travel the length of the carousel, the carousel rotates a quarter turn. From the perspective of someone in the non-rotating frame, the ball has gone straight, and the carousel has turned (middle panel). From the perspective of someone in the rotating frame, the carousel remains stationary, and the ball has curved to the right (right panel).

person sitting at point 3 and thrown toward point 1. Its speed is such that it reaches the other side of the carousel in the amount of time it takes for the carousel to rotate a quarter turn. The position of the ball after it reaches the other side is shown in the right-hand side of the figure. Notice that in inertial space (a non-rotating frame of reference), the carousel has made a quarter turn, while the ball has traveled a straight path upwards. Since the carousel has turned, the ball finishes at point 2. In the rotating frame, however (the perspective of a person *on* the carousel), the carousel appears stationary, so the ball appears to have curved right, from point 1 to point 2. This tendency to veer to the right is known as the *Coriolis force*.

To determine an equation of motion for m_3 *in the rotating frame*, one then needs to sum all of the forces (real and fictitious) acting on m_3 and apply them to Newton's second law, i.e.,

$$\Sigma \vec{F} = m_3 \ddot{\vec{r}}_3. \tag{4.1}$$

Because m_3 feels the gravitational effects of m_1 and m_2 and two fictitious forces (centrifugal and Coriolis), there are four total forces to consider when finding the equation of motion for m_3.

The equation of motion for m_3 does not have a general analytical solution (i.e., one that can be expressed with a mathematical formula). However, wherever the sum of the forces, $\Sigma \vec{F}$, is equal to zero, m_3 remains stationary in the rotating frame. In other words, at these points, m_3 has no motion relative to the other two masses. There are five points for which $\Sigma \vec{F} = 0$, first found by Jean-Louis Lagrange (c. 1700s), that have therefore come to be known as *Lagrange points*. The locations of the five Lagrange points are show in Figure 4.2.

In the next section, we'll do an approximate derivation of the location of one Lagrange point, L_1, which is straightforward (although certainly not trivial) to calculate.

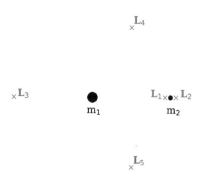

Figure 4.2. Locations of the five Lagrange points (marked with x's and labeled L_1 through L_5) when m_1 is 1000 times more massive than m_2 (for example, the Sun and Jupiter). Remember that the Lagrange points are measured in a frame rotating along with m_2. Note that the sizes of the masses are NOT to scale with the Lagrange point distances (the sizes of the objects would need to be much smaller).

Figure 4.3. Setup for calculating the location of L_1.

4.1.2 Example of Finding a Lagrange Point: L_1

L_1 is the Lagrange point that sits between the two larger masses, on a line connecting the two, as shown in Figure 4.2. To find L_1, we'll assume the forces on the smallest mass, m_3, sum to zero (remembering that we are in the rotating frame), and use the force balance equation to solve for the location of the mass. (This derivation is adapted from Stern 2016.)

At this location, m_3 will experience a gravitational force pointing to the left caused by m_1 ($F_{31,G}$), a gravitational force pointing to the right caused by m_2 ($F_{32,G}$), and a centrifugal force (F_{cent}) pointing to the right caused by the rotation of the frame around m_1. The centrifugal force is equal in magnitude (but opposite in direction) to the centripetal acceleration, mv^2/r, where r is the distance of the object from the center of the rotating frame and v is the tangential speed of the rotating frame at that r. There is actually no Coriolis force, as the Coriolis force is zero if an object is not moving, and, by definition, we are finding a point where m_3 is stationary.

Let's define r_{12} as the distance between m_1 and m_2, r_{13} as the distance between m_1 and m_3, and r_{23} as the distance between m_2 and m_3, as shown in Figure 4.3. In order for the sum of the forces on m_3 to be zero, the forces in either direction must be equal in magnitude:

$$F_{31,G} = F_{32,G} + F_{\text{cent}} \tag{4.2}$$

$$\frac{Gm_1m_3}{r_{13}^2} = \frac{Gm_2m_3}{r_{23}^2} + \frac{m_3v_3^2}{r_{13}}. \tag{4.3}$$

The velocity of the rotating frame at r_{13} is:

$$v_3 = \frac{2\pi r_{13}}{P} \tag{4.4}$$

or if we square both sides:

$$v_3^2 = \frac{4\pi^2 r_{13}^2}{P^2}, \tag{4.5}$$

where P is the period of the orbit of m_2 around m_1. The period of the orbit is given by Kepler's third law (Equation (3.49)),

$$P^2 = \frac{4\pi^2}{Gm_1}r_{12}^3 \tag{4.6}$$

so

$$v_3^2 = 4\pi^2 r_{13}^2 \frac{Gm_1}{4\pi^2}\frac{1}{r_{12}^3} \tag{4.7}$$

$$= \frac{r_{13}^2}{r_{12}^3}Gm_1. \tag{4.8}$$

We can then rewrite Equation (4.3) as:

$$\frac{Gm_1m_3}{r_{13}^2} = \frac{Gm_2m_3}{r_{23}^2} + \frac{m_3m_1Gr_{13}}{r_{12}^3} \tag{4.9}$$

$$\frac{m_1}{r_{13}^2} = \frac{m_2}{r_{23}^2} + \frac{m_1r_{13}}{r_{12}^3}. \tag{4.10}$$

If we divide both sides by r_{13}, we find:

$$\frac{m_1}{r_{13}^3} = \frac{m_2}{r_{23}^2 r_{13}} + \frac{m_1}{r_{12}^3}. \tag{4.11}$$

At this point, we'll begin to set up the equation to take advantage of the fact that $r_{23} \ll r_{12}$, using an approximation known as a *Taylor polynomial expansion* to simplify the expression. By doing this, we implicitly assume that $m_2 \ll m_1$, so that in order for the sum of forces to equal zero, L_1 would have to be pretty close to m_2. First, we note that $r_{13} = r_{12} - r_{23}$ or $r_{13} = r_{12}(1 - \frac{r_{23}}{r_{12}})$. For simplicity, we'll also define a variable $z = \frac{r_{23}}{r_{12}}$, such that $r_{13} = r_{12}(1 - z)$. Since $r_{23} \ll r_{12}$, $z \ll 1$. Then we can rewrite Equation (4.11) as:

$$\frac{m_1}{r_{12}^3(1-z)^3} = \frac{m_2}{r_{23}^2 r_{12}(1-z)} + \frac{m_1}{r_{12}^3}. \tag{4.12}$$

Now we make use of the Taylor polynomial expansion, which is a way of expressing any function as an infinite sum of polynomial terms The neat thing is that while an infinite number of terms may be required to exactly reproduce the function, one or two terms may do a pretty good job of approximating that function. The Taylor expansion of a function $f(x)$ about some point a is given by the infinite series:

$$f(x) = f(a) + \frac{f'(a)}{1!}(x-a) + \frac{f''(a)}{2!}(x-a)^2 + \cdots \tag{4.13}$$

In this case, since our expression above has both $\frac{1}{(1-z)^3}$ and $\frac{1}{(1-z)}$ let's consider a function of the form $f(x) = \frac{1}{(1-x)^n}$, and, consider the case where x is very small. In other words, we are interested in approximating the function near 0, so we substitute 0 in for a. (This is a special case of the Taylor expansion known as a Maclaurin series.) Let's compute the first two terms:

$$f(0) = \frac{1}{(1-0)^n} = 1 \tag{4.14}$$

$$f'(x) = n(1-x)^{n-1} \tag{4.15}$$

$$f'(0) = n \tag{4.16}$$

$$\frac{f'(0)}{1!}(x-0) = nx. \tag{4.17}$$

Therefore, $\frac{1}{(1-x)^n} \approx 1 + nx$ if x is close to 0 ($\ll 1$). (You can try this approximation out to see how well it does. How small does x have to be for the approximation to work well?) If we substitute these approximations into our expression, we'll find:

$$\frac{m_1(1+3z)}{r_{12}^3} \approx \frac{m_2(1+z)}{r_{23}^2 r_{12}} + \frac{m_1}{r_{12}^3} \tag{4.18}$$

or

$$\frac{m_1(1+3z)}{r_{12}^3} \approx \frac{m_2(1+z)}{z^2 r_{12}^3} + \frac{m_1}{r_{12}^3}. \tag{4.19}$$

If we divide both sides by m_1 and multiply both sides by r_{12}^3, we find:

$$1 + 3z \approx \frac{m_2}{m_1}\frac{1+z}{z^2} + 1 \tag{4.20}$$

$$3z \approx \frac{m_2}{m_1}\frac{1+z}{z^2} \tag{4.21}$$

$$3z^3 \approx \frac{m_2}{m_1}(1+z). \tag{4.22}$$

Since $z \ll 1$ we can can let $1 + z \approx 1$ and we find:

$$z \approx \left(\frac{m_2}{3m_1}\right)^{1/3} \tag{4.23}$$

$$\frac{r_{23}}{r_{12}} \approx \left(\frac{m_2}{3m_1}\right)^{1/3} \tag{4.24}$$

$$r_{23} \approx r_{12}\left(\frac{m_2}{3m_1}\right)^{1/3}. \tag{4.25}$$

Let's consider a practical example. At what distance from the Earth should we place a satellite if we'd like it to sit in the L_1 Lagrange point for the Earth–Moon system? In this case,

$$m_1 = m_\oplus = 6 \times 10^{24} \text{ kg}, \tag{4.26}$$

$$m_2 = m_\leftmoon = 7 \times 10^{22} \text{ kg}, \tag{4.27}$$

and

$$\left(\frac{m_2}{3m_1}\right)^{1/3} \approx 0.16. \tag{4.28}$$

So $r_{23} = 0.16 r_{12}$. Since the Earth–Moon distance is about 384,000 km, the satellite should be placed about 61,000 km from the Moon.

This radius also defines the boundary between when a satellite would orbit m_2 and when it would orbit m_1. The sphere of radius r_{23} surrounding m_2 is known as the *Hill sphere*. Within this radius, a satellite would have an orbit around m_2; therefore, the Hill sphere represents the sphere of gravitational influence around m_2.

If we were to repeat this exercise for all of the Lagrange points, we would find that L_2 is located at approximately the same distance from m_2 as L_1, but in the opposite direction. L_3 is located at approximately the same distance from m_1 as m_2, but opposite the location of m_2. And, the lines connecting L_4 and L_5 to m_1 and m_2 form two equilateral triangles.

4.1.3 Lagrange Points and Stability

Since there is no net force on m_3 at the Lagrange points, these are points of *equilibrium*. However, you may recall from your physics classes that there are both

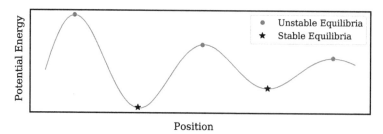

Figure 4.4. Visualization of points of stable and unstable equilibrium.

stable and unstable equilibria. For example, if you imagine a roller coaster shaped as in Figure 4.4; the peaks and dips in the roller coaster are both locations where a roller coaster can feel zero net force, and be in equilibrium. However, the lowest points are stable equilibria—if the roller coaster were given a slight push, it would leave the equilibrium point, but soon return, and, in fact, oscillate about the equilibrium point. If, on the other hand, the roller coaster is sitting on top of a peak and receives a push, it will fall precipitously off of the peak, never to return. The peaks, therefore, are unstable equilibria.

We could also interpret Figure 4.4 as a plot of the potential energy of the roller coaster car. Equilibrium points exist where the slope of the potential energy curve is zero. Stable equilibria have a positive second derivative of potential energy, while unstable equilibria have a negative second derivative of potential energy.

Similarly, there are both stable and unstable Lagrange points. At the unstable Lagrange points, L_1, L_2, and L_3, an object receiving a "kick" will leave the Lagrange point and not return. At the stable Lagrange points, L_4 and L_5, an object receiving a "kick" will return to the Lagrange point, and, in fact, oscillate (or, more precisely, orbit) about it. Orbits about L_4 or L_5 are known as "tadpole orbits" due to their shape. However, some objects end up orbiting around both L_4 and L_5, a phenomenon known as a "horseshoe" orbit.

4.1.4 Lagrange Points in the Solar System

Many objects in the solar system are found near the stable (L_4 and L_5) Lagrange points. *Trojan asteroids* are located near the Sun–Jupiter stable Lagrange points. At least two asteroids orbit around the Sun–Earth Lagrange points, and other small objects are found orbiting nearly every stable Sun–planet Lagrange point in the solar system.

Lagrange points are also convenient locations to place human-made satellites. Since the sum of the forces on the satellites would be zero, they remain in a fixed position relative to the Earth (or, if they are somewhat off from where they should be, small rocket boosts can easily bring them back near the equilibrium point). For example, the Herschel Space Observatory and the Gaia satellite reside in the Earth–Sun L_2 point, and the upcoming James Webb Space Telescope (JWST) will be launched to that point as well. At this location, JWST will always be aligned with the Earth, allowing it to communicate its results to Earth-based antennae. You might imagine the Earth could be used to block the Sun's light, but actually JWST will be slightly misaligned with Earth's orbital plane, so as to receive and use solar energy. However, large Sun shields are required to keep the instruments cool. As L_2 is $\sim 10^6$ km from Earth, servicing missions and instrument upgrades to JWST will not be possible (like they were for the Hubble Space Telescope).

4.2 Mean Motion Resonance

A resonance occurs when a driving force has just the right frequency to cause significant changes in the amplitude of an oscillation. In our everyday experience, a resonance occurs when we push someone on a swing at just the right frequency for them to swing ever higher. Similarly, planetary bodies can experience orbital resonances known as *mean motion resonances* (MMRs) if they receive periodic "kicks" at intervals that are integer or half-integer multiples of their orbital periods. This can occur when two objects orbiting a third object have periods that are an integer ratio. In such cases, the two objects will meet repeatedly at the same orbital locations, and the gravitational interactions between the two bodies will tend to add up over time and cause large overall orbital changes. As an example, in Figure 4.5, masses 2 and 3 (orbiting mass 1) might have $P_3/P_2 = 2$, $P_3/P_2 = 3/2$, $P_3/P_2 = 3$, etc. Any integer period ratio N/M is called an N:M resonance. The three resonance configurations just listed would be called a 2:1 resonance, 3:2 resonance, and 3:1 resonance, in that order.

The precise consequence of the MMR depends on the details of the system, and must be determined with detailed analysis or by empirical study. In the Galilean satellite system, Ganymede, Europa and Io have a 4:2:1 period ratio for their orbits around Jupiter. Ultimately, this results in a transfer of energy from Jupiter to the interiors of Io and Europa. The resonance locks the satellites into slightly non-circular orbits, in which the moons get flexed during each orbit. This flexing heats their interiors, making Io the most volcanically active body in the solar system (see Figure 4.6), and turning the icy interior of Europa into a potentially habitable ocean.

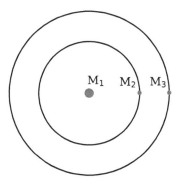

Figure 4.5. Visualization of 2 masses (M_2 and M_3) orbiting a much larger third mass (M_1) in a 2:1 mean motion resonance. Each time M_3 orbits M_1 once, M_2 orbits twice; therefore, M_2 and M_3 encounter each other (once per orbit of M_3) in the same configuration each time.

Figure 4.6. Io's surface imaged with the Galileo spacecraft, showing an erupting volcanic plume. Io's interior is heated via its mean motion resonances with Europa and Ganymede. Image Source: NASA/JPL/DLR.

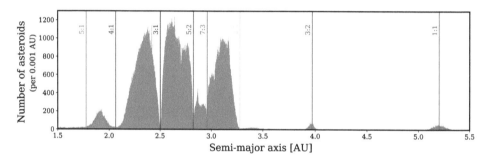

Figure 4.7. Semimajor axis distribution of asteroids (data from JPL 2020). Vertical lines mark mean motion resonances with Jupiter. Some of these resonances are destructive, and form gaps; in the asteroid belt, these are known as the Kirkwood gaps. Some of these resonances are protective, and cause accumulations of objects. Objects in the 3:2 resonance with Jupiter are known as Hildas; objects in the 1:1 resonance (specifically, in the L_4 and L_5 Lagrange points) are known as Trojans.

In the asteroid belt, MMRs with Jupiter are unstable for asteroid orbits, and over time any asteroids in these regions get removed. Today's asteroid belt shows a distinct lack of asteroids at or near the MMRs with Jupiter, as shown in Figure 4.7. Finally, in the Kuiper Belt (the region of objects orbiting beyond Neptune, of which Pluto is one member), MMRs with Neptune are protective. The distribution of objects in the Kuiper Belt shows a surplus at or near the Neptune MMRs (see Figure 4.8). Pluto itself is in a 3:2 resonance with Neptune, allowing it to remain in a stable orbit that actually crosses the orbit of Neptune. Objects in the 3:2 resonance have come to be known as Plutinos.

4.3 Tides and Tidal Synchronization

In our previous discussions of dynamics, we have treated all objects as point masses and have ignored the fact that they have finite sizes. *Tidal forces* are the forces that bodies experience due to the fact that the strength of the gravitational force is not constant throughout the object. Aside from causing the ocean tides that we

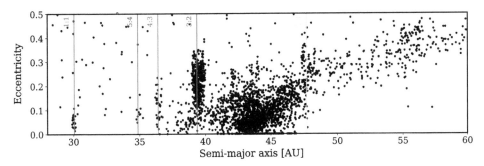

Figure 4.8. Eccentricity versus semimajor axis for objects in the Kuiper Belt (data from JPL 2020). Vertical lines mark mean motion resonances with Neptune. These resonances are protective, allowing for accumulations of objects, and for survival even for high eccentricities (which would otherwise tend to cause collisions with Neptune). Objects in the 1:1 resonance are Neptune Trojans, in the L_4 and L_5 Lagrange points.

Not to scale

Figure 4.9. Setup for derivation of tidal forces on the Moon from the Earth. For the purposes of calculating the tidal forces *on the Moon*, we ignore the physical size of the Earth and treat it as a point source. Note that neither the sizes nor the distances are to scale in this figure.

experience here on Earth, tidal forces cause the synchronization of spin and orbital periods, and a concurrent release of tidal energy in the interior of the affected bodies.

4.3.1 Magnitude of the Tidal Force

Any real object experiencing a gravitational force will also experience tidal forces, due to the fact that the gravitational force is not constant throughout the object. Since different sides of the body experience different magnitudes of gravitational force, the body itself feels as if it's being pulled apart. This pulling apart is known as the tidal force. The strength of the tidal force depends on the mass of the body producing the gravitational force, the distance between the objects, and the size of the object experiencing the force, the rough details of which we will derive here. We are most familiar with the tidal forces the Earth experiences from the Moon; however, the Earth also experiences tidal forces from the Sun, and the Moon itself experiences tidal forces from the Earth. It is this last situation that we will consider here, so that we can discuss *tidal synchronization* in the next section. However, this derivation can be easily generalized to any pair of bodies.

As shown in Figure 4.9; we consider a system consisting of the Earth and the Moon, separated by a distance r. Although the Earth has a finite size and thus experiences tidal forces from the Moon, we will ignore these forces, and focus instead on the tidal forces that the Moon experiences from the Earth. We'll let the radius of the Moon be R, the mass of the Earth be $m_\oplus$, and consider the forces on a point mass m. If that point mass were located where the center of the Moon is, it would experience a force:

$$F_0 = \frac{Gm_\oplus m}{r^2}.$$

If, instead, the point mass were on the side of the Moon closest to the Earth, it would experience a force:

$$F_{\text{in}} = \frac{Gm_\oplus m}{(r - R)^2}.$$

As we would expect, $F_{\text{in}} > F_0$. Since the Moon is essentially undergoing uniform circular orbital motion, the gravitational force at the center of the Moon would need to be exactly enough to keep a point mass in perpetual circular motion, i.e., $F_0 = \frac{mv^2}{r}$. But for the part of the Moon closer to the Earth, the force $F_{\text{in}} > \frac{mv^2}{r}$, and this part of the Moon is pulled slightly toward the Earth. Similarly, the part of the Moon farther from the Earth experiences a force $F_{\text{out}} < \frac{mv^2}{r}$, and the net effect is for the more distant part of the Moon to get pulled away from the Earth. Thus, although the *absolute* gravitational forces experienced by the Moon all point toward the Earth, *relative* to the force experienced at the center of the body (which is exactly the right amount to cause orbital motion), the *differential* forces on the Moon are inwards on the inner edge, and outwards on the outer edge. These differential forces are known as tidal forces. This concept is displayed in Figure 4.10.

The magnitude of the tidal force is approximately given by the difference $F_{\text{in}} - F_0$:

$$F_{\text{in}} - F_0 = \frac{Gm_\oplus m}{(r - R)^2} - \frac{Gm_\oplus m}{r^2}. \tag{4.29}$$

At this point, we'll rewrite the expression in a clever way, so that we can ultimately take advantage, again, of a Taylor polynomial expansion:

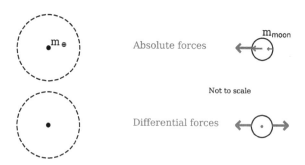

Figure 4.10. Top: Visualization of absolute gravitational forces experienced by the Moon from the Earth. Bottom: Visualization of differential (tidal) gravitational forces experienced by the Moon. (Forces NOT to scale.)

$$F_{\text{in}} - F_0 = \frac{Gm_\oplus m}{(r(1 - \frac{R}{r}))^2} - \frac{Gm_\oplus m}{r^2} \tag{4.30}$$

$$= \frac{Gm_\oplus m}{r^2(1 - \frac{R}{r})^2} - \frac{Gm_\oplus m}{r^2}. \tag{4.31}$$

In Section 4.1.2, we found that any function of the form $\frac{1}{(1-x)^n}$ can be approximated by $1 + nx$ if $x \ll 1$. Since $R \ll r$ (i.e., the size of the Moon is much smaller than the Earth–Moon distance), $R/r \ll 1$, and we can approximate $\frac{1}{\left(1 - \frac{R}{r}\right)^2}$ as $1 + \frac{2R}{r}$. Making this substitution, we find:

$$F_{\text{in}} - F_0 \approx \frac{Gm_\oplus m}{r^2}\left(1 + 2\frac{R}{r}\right) - \frac{Gm_\oplus m}{r^2} \tag{4.32}$$

$$\approx \frac{2Gm_\oplus mR}{r^3}. \tag{4.33}$$

Because we were considering the force on an arbitrary point mass, m, it makes more sense to consider the tidal force per unit mass,

$$F_T = \frac{2Gm_\oplus R}{r^3}. \tag{4.34}$$

As we discussed earlier, the tidal force depends on the mass of the body providing the gravitational force (in this case, $m_\oplus$), the size of the body experiencing the force (R), and the distance between them (r). However, it's interesting to note that there is a cubed dependence on the distance between the objects, meaning that tidal forces decrease rapidly in magnitude as the objects get farther apart.

Considering the tidal force, we might also ask—can the force ever be large enough to pull a body apart? The answer to this is yes. As a homework problem, you will derive the so-called Roche limit—the closest distance a moon can get to the body it's orbiting before it is torn apart by tidal forces. It's believed that rings in the solar system are composed of moons that were torn apart by tidal forces.

4.3.2 Tidal Synchronization

In the previous section, we discussed how the *differential* tidal forces pull one end of the Moon toward the Earth, and the other end away from the Earth. It can sometimes be helpful to picture the tidal forces as producing *tidal bulges* caused by the tidal forces, or, in other words, an elongation of the object along the direction of the tidal forces. The Moon is both orbiting the Earth, and rotating on its own axis. If the Moon's rotation rate is exactly equal to its orbital angular speed, the so-called tidal bulges formed by these forces always remain aligned with the Earth–Moon line. This idea is shown in the top of Figure 4.11, with the Moon's tidal bulging here being extremely exaggerated.

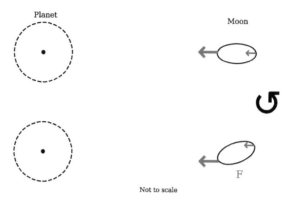

Not to scale

Figure 4.11. Exaggerated visualizations of aligned (top) and misaligned (bottom) tidal bulges. (In reality, the angle of misalignment, and the size of the tidal bulges would be much smaller.) The black circular arrow shows the direction of the Moon's orbit and rotation. The misaligned bulge would result if the rotational angular speed is slightly faster than the orbital angular speed. Straight blue arrows show exaggerated gravitational forces felt by the Moon from the planet. For the aligned bulge, there is no net torque on the Moon. For the misaligned bulge, the difference in forces on the two ends of the bulge result in a torque that acts opposite to the direction of rotation, slowing down the Moon's rotation.

However, what happens if the rotation rate and orbital angular speed are not the same? Let's consider what happens if the rotation rate is faster than the orbital angular speed. In this case, the relatively fast rotation rate will cause the tidal bulge to misalign slightly with the Earth–Moon line, as demonstrated (in an exaggerated sense) in the bottom of Figure 4.11. One part of the tidal bulge is slightly closer to the Earth and experiences a slightly enhanced gravitational force. The other side of the bulge is slightly farther from the Earth and experiences a slightly diminished gravitational force. The net effect is that the Moon will experience a slight torque in a sense that is opposite the direction of rotation. Thus, this torque will cause a slight reduction in the Moon's rotation rate. After a sufficient amount of time has passed, the continuous reduction in rotation rate will cause the rotation rate and orbital angular speed to match, the tidal bulge will remain aligned with the Earth–Moon line, and no torque will be experienced by the Moon. At this point, we say that the Moon is *tidally locked*. Its rotation period is equal to its orbital period—it has achieved *spin–orbit synchronization*, and observers on Earth always see the same side of the Moon.

Tidal synchronization has occurred throughout the solar system, and most pairs of bodies are evolving toward this state. For the Earth–Moon system, while the Moon has achieved *spin–orbit synchronization*, the Earth is currently spinning faster than the Moon orbits the Earth. Thus, the tidal bulges on the Earth from the Moon are misaligned with the Earth–Moon line, and the Earth is slowly evolving toward a state where the Earth's rotation period will match the Moon's orbital period. Elsewhere in the solar system, Phobos and Deimos are in synchronous rotation with Mars, and the Galilean satellites are in synchronous rotation with Jupiter. Pluto and its moon Charon represent the future state of the Earth–Moon system—Pluto's rotation, Charon's rotation, and Charon's orbit all have the same period.

4.3.3 Ocean Tides on Earth

Tidal forces on the Earth caused by the Moon are responsible for the ocean tides on Earth. Because the oceans are able to move freely, the tidal forces cause a slight migration of water toward the tidal bulges—the cause of high tides. 90° away from these points, locations experience low tides. As the Earth rotates, the locations on Earth experiencing high and low tides slowly change. Since both the near and far side of the Earth experience tidal bulging, every 24 hr a given location experiences *two* high tides and *two* low tides.

The rapid decrease in tidal force with distance (remember that $F_T \propto \frac{1}{r^3}$) is the reason why the tides on Earth are mostly caused by the Moon, rather than the Sun. The Sun is much more massive than the Moon, but it is also much farther away. In fact, we can estimate the ratio of their effects by calculating the ratio of the tidal force on Earth due to the Sun to the tidal force on Earth due to the Moon:

$$\frac{M_\odot}{m_{\text{Moon}}} \left(\frac{r_{\text{Earth-Moon}}}{r_{\text{Earth-Sun}}} \right)^3 = \frac{2 \times 10^{30} \text{ kg}}{7 \times 10^{22} \text{ kg}} \times \left(\frac{4 \times 10^8 \text{ m}}{2 \times 10^{11} \text{ m}} \right)^3 \quad (4.35)$$

$$\approx 0.23. \quad (4.36)$$

Therefore, the tidal forces on the Earth caused by the Sun are weaker than those from the Moon, although the effect from the Sun is not insignificant.

Although the Moon has a stronger tidal effect on the Earth than the Sun, the Sun does produce a non-negligible effect on Earth's ocean tides. When the Sun is aligned with the Earth and Moon, the Earth experiences stronger tides known as spring tides (see Figure 4.12). When the Sun is maximally misaligned with the Earth and Moon, the Earth experiences weaker tides known as neap tides. The Earth experiences neap and spring tides twice per lunar orbit. Notice that during spring tides, observers on Earth will also experience either a new moon or full moon, while during neap tides, observers on Earth will see a first or third quarter moon (in which the Moon looks half illuminated).

4.4 Important Terms

- circular restricted three body problem;
- fictitious forces;
- Coriolis force;
- Lagrange points;
- Taylor polynomial expansion;
- Hill sphere;
- stable and unstable equilibrium;
- Trojan asteroids;
- mean motion resonance;
- tidal force;
- tidal bulges;
- tidal synchronization/spin–orbit synchronization.

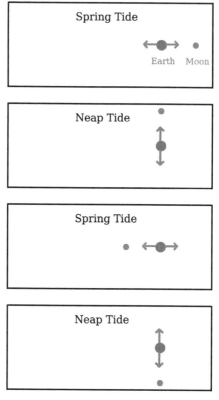

Figure 4.12. Visualization of spring and neap tides (distances not to scale). Yellow, blue, and gray circles represent the relative positions of the Sun, Earth, and Moon, in that order, at four different times during a lunar orbit. Small yellow and larger gray arrows show the directions of the solar and lunar tidal forces, respectively.

4.5 Chapter 4 Homework Questions

1. The Roche limit:

   The Roche limit is the distance from a more massive body within which tidal forces would rip apart any less massive orbiting body. In this problem, you will do a rough derivation of the Roche limit for a planet. To begin, make sure you recall that for some particle of mass m on the surface of a moon orbiting a planet of mass m_p, the magnitude of the tidal force is approximately $F_{\text{Tidal}} = \dfrac{2RGm_pm}{r^3}$ where R is the radius of the Moon and r is the distance between the Moon and the planet.

   (a) The tidal force tries to rip the Moon apart, but the Moon's own gravity will try to hold it together. Write down an expression for the magnitude of the gravitational force felt by a particle of mass m on the surface of the Moon (i.e., at a distance of R from the center of the Moon) due to the Moon's own gravity. Use the symbol m_m to represent the mass of the Moon. Make sure you keep all of your symbols straight! (Note: You are completely ignoring the tidal force for this part.)

(b) By setting the tidal force exerted by the planet equal to the gravitational force exerted by the Moon itself, find an expression for the distance, r, from the planet, at which these two forces are equal. Beyond this distance, the gravitational forces "win"; inside this radius, the tidal forces "win". Make sure your answer makes sense by checking the units and checking that it does what you expect. (What happens to r if you increase the mass of the planet? What happens to r if you increase the mass of the Moon? What happens to r if you increase R? Are these what you expect?)

(c) The equation you derived is not in a very useful form, since it depends on the size of the Moon. Instead, we want a general expression for how far *any* moon would have to be from a planet to survive tidal forces. Take your expression for r and replace each mass with the equation for mass $M = 4/3\pi R^3 \rho$ where ρ is density and R the radius of the body. To keep your variables straight, make sure you use subscripts to differentiate the Moon and planet. Simplify to find an expression for r that depends only on the relative densities of the Moon and planet and the radius of the *planet*. You have just found the definition of the Roche limit.

(d) Our simplified derivation gets the variables right, but is off by a small constant (due in part to the fact that the Moon gets tidally stretched). The correct expression for the Roche limit is $r \approx 2.5 \times R_p \left(\frac{\rho_p}{\rho_m} \right)^{1/3}$. Using this expression, estimate the Roche limit for Saturn. For the Moon, assume it's an icy mixture with a density of $1000 \, \text{kg m}^{-3}$.

(e) Compare your answer to the radius of Saturn, the orbital radius of Saturn's A-ring—the most prominent outer ring, as well as to the orbital radius of Janus and Epimetheus—moons that may be currently falling apart to create a new ring. Discuss your answer.

(f) Estimate the Roche limit for the Earth, and compare to the location of the Earth's Moon. Discuss your answer.

2. Tidal synchronization:

The same side of the Moon always faces the Earth, because the Moon's orbital period and rotation period are the same. The Earth/Moon tides are slowly causing Earth's rotation to slow down, and for the Moon to move away from the Earth. Eventually, the Earth's rotation rate will be the same as the Moon's, and only half of the Earth will be able to see the Moon. In this question, you'll use the principle of conservation of angular momentum to find the final Earth–Moon distance, and the length of an Earth day.

(a) Write down an expression for the angular momentum associated with the Earth's rotation. Assume the Earth is a uniform density sphere with mass $M_\oplus$, radius $R_\oplus$, and rotation rate $\omega_\oplus$.

(b) Write down an expression for the angular momentum associated with the Moon's orbit around the Earth. Assume the Moon has a mass $m_{\leftmoon}$, an orbital semimajor axis $a_{\leftmoon}$, and orbital rate $\Omega_{\leftmoon}$.

(c) Calculate the current angular momenta associated with the Earth's rotation and the Moon's orbital motion, respectively, using the expressions you just derived. Which one is larger, and by what factor?

(d) Assuming the Moon is in a circular orbit around the Earth, derive an expression for $\Omega_{\leftmoon}$ in terms of $M_{\oplus}$, $a_{\leftmoon}$ and any relevant constants.

(e) Using your answer to the previous question, rewrite the expression for the Moon's orbital angular momentum by removing the dependence on $\Omega_{\leftmoon}$.

(f) For our next calculation, we're going to assume that the angular momentum associated with Earth's rotation will be small enough to ignore in our calculation once the Moon reaches its final orbit. Based in part on your answer to part (c), explain why this may be a reasonable assumption.

(g) Using conservation of angular momentum, and assuming that Earth's final rotational angular momentum is small enough not to matter in this calculation, find an expression for the final semimajor axis of the Moon's orbit.

(h) Calculate the ratio of the Moon's final and current positions, $a_{\leftmoon,f}/a_{\leftmoon,i}$.

(i) Calculate the final length of a day on Earth.

(j) Assess and discuss your answers.

References

NASA/JPL Solar System Dynamics Group 2020, JPL Small-Body Database Search Engine, https://ssd.jpl.nasa.gov/

National Geographic 2014, Coriolis Effect, https://http://www.youtube.com/watch?v=mPsLanVS1Q8

Stern, D. P. 2016, The Distance to the L1 Point, http://www.phy6.org/stargaze/Slagrang.htm

Introductory Notes on Planetary Science
The solar system, exoplanets and planet formation
Colette Salyk and Kevin Lewis

Chapter 5

Extrasolar Planets

In this chapter, we'll discuss the detection and characterization of extrasolar planets (also known as exoplanets)—planets orbiting stars other than the Sun.

5.1 Why Finding Exoplanets Is Difficult

If we want to find a planet, our first instinct may be to simply try to take a picture of it. Indeed, the planets in the solar system were all detected via "direct imaging". However, direct imaging of extrasolar planets turns out to be difficult, because extrasolar planets are both faint and relatively close to their parent star.

In Section 2.2 we discussed how the amount of light emitted by a body, and the peak of its blackbody spectrum, depend on temperature. Stars, since they are actively fusing hydrogen, have surface temperatures of thousands of Kelvin (or more), while planets have temperatures close to their equilibrium temperatures (typically 100's of Kelvin or less; recall Equation (2.15)). When trying to image a planet, therefore, we have to deal with the "glare" of the bright star it's orbiting.

Figure 5.1 shows the observed luminosity of the Sun, Earth, Jupiter, and Uranus. Note the two bumps in the planet curves. The bumps at longer wavelengths represent the thermal emission from the planets (peaking at wavelengths determined by Wien's law—Equation (2.3)) that we discussed in Chapter 2. The other, shorter wavelength, bumps are the reasons why we can see solar system planets with our own eyes—they arise from the sunlight reflected by the planet. These curves demonstrate an unfortunate fact about direct imaging—no matter what wavelength we choose, the contrast in brightness between the Sun and the planets remains high. For this reason, direct imaging of exoplanets is difficult, and was not, historically, the first successful or most efficient method for detecting exoplanets. In the next sections, we will discuss the most successful methods, and then we will revisit the technique of direct imaging and discuss the advanced techniques used to combat the difficult issue of *star/planet contrast.*

5-1

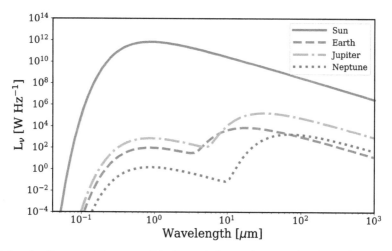

Figure 5.1. Luminosity per unit frequency of the Sun and planets, demonstrating the star/planet contrast as a function of wavelength. The short and long wavelength bumps in the planet curves correspond to reflected sunlight and blackbody emission, respectively. No matter what wavelength we choose, the Sun appears many orders of magnitude brighter than the planets. (Planetary parameters from the NASA Planetary Fact Sheet NASA 2020.)

5.2 Radial Velocity

Because planets themselves are difficult to image, we can instead try to figure out if the star is undergoing an orbital dance with an unseen exoplanet (see Figure 3.2). The radial velocity technique takes advantage of the fact that a star and planet *both* orbit their mutual center of mass.

5.2.1 Review of Doppler Shift

The *Doppler shift* can be experienced in our everyday life in the context of sound waves. If something, like an ambulance, is emitting a sound wave and moving away from us, that sound wave gets stretched out and we hear the frequency of the sound decrease (a lower pitch). If the ambulance is moving toward us, however, the sound wave gets compressed, and we hear an increase in the sound frequency (a higher pitch). Like sound, light is also a wave, and light waves get stretched or compressed if emitted by an object moving toward or away from us.

The magnitude of the Doppler shift of light is given by

$$\Delta\lambda = \frac{v_r}{c}\lambda_0 \tag{5.1}$$

where λ_0 is the wavelength of light that is emitted, v_r is the light source's velocity toward or away from the observer (also known as the *radial velocity*), c is the speed of light, and $\Delta\lambda$ is the observed shift in wavelength caused by the motion.

If a star is being orbited by an unseen planet, and is continually moving toward and away from us as it orbits the center of mass of the system, its light will get continually Doppler shifted.

5.2.2 The Magnitude of the Doppler Shift for a Planet-hosting Star

We can use our knowledge of orbital mechanics to estimate the velocity of a star being orbited by a planet, and calculate the magnitude of the observed Doppler shift. For the sake of simplicity, we will discuss the case of perfectly circular orbits, and assume that the mass of the planet (m_p) is much less than the mass of the star ($M_\star$).

For a circular orbit, the velocity of the star is given by

$$v_\star = \frac{\text{distance}}{\text{time}} \tag{5.2}$$

$$= \frac{2\pi a_\star}{P_\star} \tag{5.3}$$

where $a_\star$ is the semimajor axis of the star's orbit and $P_\star$ is the period of its orbit. In Section 3.2.3 we showed that the position of the star is always opposite that of the planet, and that the size of its orbit is diminished by a factor $m_p/M_\star$ (Equation (3.41)). Thus, the semimajor axis of the star's orbit is given by

$$a_\star = \frac{m_p}{M_\star} a_p, \tag{5.4}$$

but its period is equal to that of the planet's period, $P_\star = P_p$.

The planet's period can be related to its semimajor axis via Kepler's third law (Equation (3.49)):

$$P_p^2 = \frac{4\pi^2 a_p^3}{GM_\star} \tag{5.5}$$

$$P_\star = P_p = \frac{2\pi a_p^{3/2}}{\sqrt{GM_\star}}. \tag{5.6}$$

Substituting $a_\star$ and $P_\star$ back into Equation (5.3), we find:

$$v_\star = \frac{2\pi \frac{m_p}{M_\star} a_p}{\frac{2\pi a_p^{3/2}}{\sqrt{GM_\star}}} \tag{5.7}$$

$$= m_p \sqrt{\frac{G}{M_\star a_p}}. \tag{5.8}$$

This expression shows us what types of planets correspond with the largest stellar velocities—planets that are massive (large m_p), planets that are close to their parent stars (small a_p) and planets around lower-mass stars (small $M_\star$).

Once we know the velocity of the star, we can determine the magnitude of the observed Doppler shift. However, it's important to remember that the Doppler shift depends on the *radial* velocity of the source. This has two important consequences.

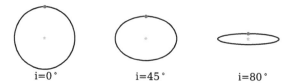

i=0°   i=45°   i=80°

Figure 5.2. Visualization of different inclinations with respect to an observer, assuming a circular orbit. (The planet and star are not to scale with the orbital size.)

First, the amount of Doppler shift depends on the orientation of the orbit with respect to the observer, known as the orbital inclination, i. The convention in astronomy is that $i = 0°$ if the orbit is viewed face on, while $i = 90°$ if the orbit is viewed perfectly edge on. A few example inclinations are shown in Figure 5.2. A perfectly face-on orbit would have no motion toward or away from the observer, and so no detectable Doppler shift. A perfectly edge-on orbit would have the maximum possible motion toward or away from the observer. In general, the more edge-on the orbit is, the larger the signal. The maximum observed radial velocity is given by the equation:

$$v_r = v_\star \sin i. \tag{5.9}$$

This behaves as we expect—$v_r = 0$ for $i = 0$ and $v_r = v_\star$ for $i = 90°$. Similarly, then, the observed Doppler shift is given by:

$$\Delta\lambda = \frac{v_\star \sin i}{c}\lambda_0 \tag{5.10}$$

$$= \frac{m_p \sin i}{c}\sqrt{\frac{G}{M_\star a_p}}\lambda_0. \tag{5.11}$$

When a Doppler shift is observed, there is no way to know the inclination of the orbit (unless a complementary detection technique is also used—see discussion in upcoming sections). Therefore, if we observe some $\Delta\lambda$, it may be the case that $i = 90°$ and that the radial velocity can be directly related to planet mass via Equation (5.11). However, it is much more likely that the orbit is not exactly edge-on. In such a case, the Doppler shift only tells us the value of $m_p \sin i$. This is referred to as the planet's *minimum mass*, since the true mass is given by $m_p = \frac{m_p \sin i}{\sin i}$ and $\sin i$ takes values between 0 and 1.

The second consequence of the fact that it is the *radial* velocity that is observed is that the observed Doppler shift changes with time as the star orbits. Even for a purely edge-on orbit, there are parts of the orbit where the star is moving directly toward or away from the observer, and parts of the orbit where the motion is purely transverse with respect to the observer, and would produce no observable Doppler shift. By convention, we consider movement away from the observer to be a positive velocity, and motion toward the observe to be a negative velocity. For an edge-on orbit, $v_r = v_\star \cos \omega t$ where ω is the angular speed and t is time. Thus, the observed radial velocity is a curve, as shown in Figure 5.3. If the orbital inclination is not zero, the full description of the radial velocity is given by $v_r = v_\star \sin i \cos \omega t$, and its maximum value is given by $v_{max} = v_\star \sin i$.

5.2.3 Observing the Doppler Shift

To actually observe the wavelength shift associated with the Doppler shift, we have to produce a spectrum of the star, and observe how it changes with time. A spectrum is a plot of the amount of light observed as a function of wavelength. Because stars have atmospheres that are cooler than their interiors, they produce spectra that are a combination of a continuous (blackbody radiation) spectrum and an absorption spectrum. The dark absorption lines are due to absorption of light by atoms or molecules in the star's atmosphere. An example stellar spectrum—in this case, for the Sun—is shown in Figure 5.4.

The spectrum shown in Figure 5.4 is produced by placing a diffraction grating (a finely-grooved piece of glass that disperses the light into its component wavelengths)[1] behind a narrow slit that confines the incoming light. While the horizontal dimension of each row in the plot represents wavelength, the vertical dimension simply shows how the spectrum varies, if at all, along the length of the slit. Assuming the source is uniform along the slit, the signal can be collapsed down along this dimension to produce a one-dimensional spectrum. Figure 5.5 shows a close-up of the solar spectrum collapsed down to its one-dimensional version. This version of the spectrum is more readily suitable for quantitative analysis than the pictorial representation above.

In order to detect a planet, stars are repeatedly observed to produce a series of spectra, and observers look for small shifts in wavelength between spectra, indicative of a Doppler shift. They then plot the Doppler shift versus time. As a homework assignment, you will calculate a typical Doppler shift due to a giant planet, and find that it is incredibly small. Therefore, the Doppler shift technique requires a few unique tricks. First, the star is observed over a large range of wavelengths, so that many absorption lines can be detected. When the positions of many lines are measured, our level of uncertainty about their average shift decreases (specifically, the uncertainty of the shift depends on the factor $1/\sqrt{N}$ where N is the number of lines measured—a result from statistics known as the central limit theorem). Second, the stars are observed in concert with some calibration source, which tells the

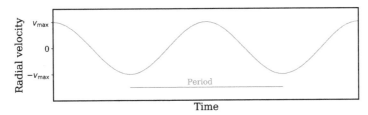

Figure 5.3. Visualization of a radial velocity curve for a star hosting a planet in a circular orbit.

---

[1] As well as a "cross-disperser", which separates the long spectrum into pieces, so they can be imaged on the same detector.

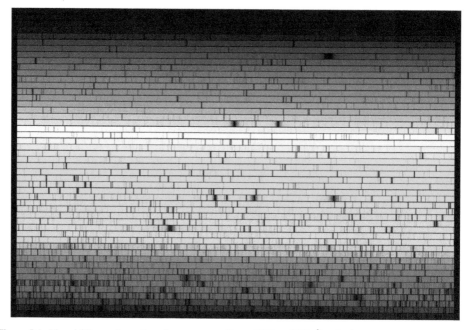

Figure 5.4. The visible portion of a solar spectrum (from 4000 to 7000 Å), obtained at the McMath–Pierce Solar Facility at the National Solar Observatory on Ioligam Doag/Kitt Peak. Wavelengths increase from left to right along each row, and from bottom to top. Dark bands are caused by absorption of photons by atoms in the Sun's atmosphere. This image is highly processed compared to the raw data one would receive at a telescope. Among other changes, colors have been added as a visual aid, and spectra have been straightened out and stacked on top of each other. Image source: N. A. Sharp, NOIRLab/NSO/Kitt Peak FTS/AURA/NSF.

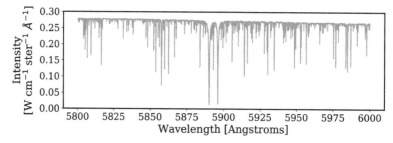

Figure 5.5. A collapsed, one-dimensional spectrum of the Sun from 5800 to 6000 Å, showing the many absorption bands arising from atoms in the Sun's atmosphere (Brault & Neckel 1987). The two deep lines at 5 889.950 and 5 895.924 Å are due to absorption from sodium, and are known as the "Sodium D lines". If you've ever placed sodium in a Bunsen burner in Chemistry class and seen the bright yellow flame that results, you've seen *emission* in these same bands.

observer exactly what wavelengths they're observing. As one example of this calibration, some of the first radial velocity detections used a small capsule of pure iodine gas, placed just beyond the detector; as the light from the star passed through the capsule, the iodine absorption lines became superimposed on the stellar spectrum, providing a means to accurately determine the wavelengths of the spectrum. While the star moved under the gravitational influence of an unseen

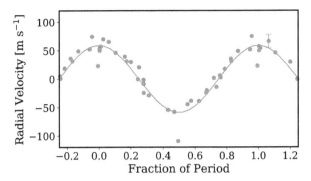

Figure 5.6. Original observed radial velocity curve for 51 Peg b—the first detected exoplanet around a Sun-like star (Mayor & Queloz 1995). The *y*-axis shows the radial velocity signal, while the *x*-axis shows the fraction of 51 Peg b's 4.2 day orbital period. One representative ~12 m s^{-1} error bar is shown. The orange sine curve has an amplitude of 59 m s^{-1}. This detection resulted in the 2019 Nobel Prize in Physics.

planet, the iodine remained stationary inside of the telescope, so the stellar absorption lines appeared to move relative to the iodine absorption lines. Finally, modern instruments are designed to have as few moving parts as possible, so that their wavelengths remain the same over long periods of time.

Figure 5.6 shows the radial velocity curve of the first detected exoplanet around a Sun-like star, 51 Pegasus b (51 Peg b, for short)—work which resulted in the Nobel prize in Physics in 2019 (Mayor & Queloz 1995). Each point represents the Doppler shift measured in a single spectrum of the star, 51 Peg. The star was observed over a period of 18 months, but the observations were overlapped (a process known as phase folding) to increase the clarity of the data. Essentially, once the period of the signal has been determined, one can take all data and wrap it back on itself every period, so that the maxima and minima of the curves all align. The *x*-axis of the curve can then be expressed as a fraction of a period, rather than time. 51 Peg b was found to have a 4.2 day orbit—much, much shorter than an Earth year, and even shorter than Mercury's 88 day year—and this for a planet with a minimum mass of 0.5 Jupiter masses. We will return to this surprising result when we discuss the characterization of exoplanets.

5.3 Astrometry

Astrometry refers to the precise measurement of stellar positions. The astrometric technique is also an attempt to detect the star's "dance" around the center of mass, but by directly measuring its motion in the sky, rather than indirectly measuring motion of the star toward or away from us via the Doppler shift. For a circular orbit, the physical offset of the star relative to the center of mass of the system is given by Equation (5.4), $a_\star = \frac{m_p}{M_\star} a_p$. For a face-on orbit, the observed *angular* diameter of the star's orbit (i.e., how large the orbit actually appears in the sky to an observer on Earth) is then given by:

$$\theta_\star [\text{radians}] = \frac{2m_p a_p}{M_\star d} \tag{5.12}$$

$$\theta_\star[°] = \frac{2m_p a_p}{M_\star d} \frac{360}{2\pi} \tag{5.13}$$

where d is the distance to the star from Earth.

In practice, $\theta_\star$ can be quite small, and difficult to detect. For example, using the distance to 51 Peg of 5×10^{17} m (Prusti & de Bruijne et al. 2016; Gaia Collaboration et al. 2018), planetary and stellar masses of $\sim 1 \times 10^{27}$ kg and $\sim 2 \times 10^{30}$ kg, respectively (Mayor & Queloz 1995), and a planetary semimajor axis of 0.05 au (Mayor & Queloz 1995), 51 Peg b would produce an astrometric signal of

$$\theta_\star[°] = \frac{2(1 \times 10^{27} \text{ kg})(0.05 \text{ au})}{(2 \times 10^{30} \text{ kg})(5 \times 10^{17} \text{ m})} \frac{360}{2\pi} \tag{5.14}$$

$$= \frac{2(1 \times 10^{27} \text{ kg})(0.05 \times 1.5 \times 10^{11} \text{ m})}{(2 \times 10^{30} \text{ kg})(5 \times 10^{17} \text{ m})} \frac{360}{2\pi} \tag{5.15}$$

$$\approx 10^{-9}. \tag{5.16}$$

10^{-9} degrees certainly sounds very small in terms of everyday experience, but what about if we use a telescope that can make very precise angular measurements? Astronomers often need to discuss very small angles, so they divide each degree of angle into 60 arcmin, and each arcminute into 60 arcsec, meaning that there are 3600 arcsec per degree. If we convert our astrometric signal to arcseconds, we multiply 10^{-9} by 3600 to find that the signal is about 4×10^{-6} arcsec. Today's most powerful (roughly 10 m diameter) telescopes, observing at visible wavelengths, and after correcting for any blurring by Earth's atmosphere, produce stellar images only as small as about 0.02 arcsec!![2] So observing the astrometric signal caused by 51 Peg b is not feasible with current ground-based telescopes.

While astrometric detection of planets remains difficult, the signal is larger for more massive planets far from their parent stars, for less massive stars, and for the most nearby stars. One telescope that is particularly well-suited to detect astrometric signals of exoplanets is the European Space Agency's Gaia spacecraft (Prusti & de Bruijne et al. 2016). It was specifically designed to measure precise positions of stars in the entire sky; in fact, the best precision it can achieve (for bright stars) is about 10 micro-arcsec. The spacecraft was launched in 2013 and is scheduled to remain in operation until at least 2022. It is expected to discover thousands of nearby exoplanets. Note that if a full orbit is detected, the astrometric technique provides the inclination of the stellar orbit. As the inclination of the orbit increases, the orbit looks more and more elliptical to an observer near Earth. However, we can also measure the speed of the star at each point in its orbit, which we know will vary with position if the orbit is eccentric, according to Kepler's second law. Thus the true

---

[2] Technically speaking, the precision on the star's location is somewhat better than this, as the star's position can be determined more precisely than its width—another result of the statistical concept known as the central limit theorem.

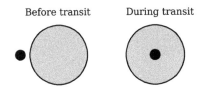

Figure 5.7. Setup for calculation of transit dip strength.

mass (not the minimum mass) of the unseen companion can be determined with this method.

5.4 Transits

The transit detection technique takes advantage of a very simple idea—if a planet passes in front of a star, it will block part of the surface of the star, causing a slight decrease in the amount of light reaching an observer. The passage of the planet in front of the star is called a *transit*, and the plot of light from the star versus time is known as a *lightcurve*. Although the idea is simple, the actual detection of planets (via a transit) is challenging for several reasons. First, planets have much smaller radii than stars, so the percent decrease in light from the star is quite small. Second, the planet and star need to appear nearly perfectly aligned with respect to the observer if the planet is to block light from the star—a rare situation since orbital inclinations should be randomly distributed. In addition, the star must be observed as the planet is passing in front of the star, but the planet spends only a small fraction of its orbit in this configuration. Therefore, in order to detect an exoplanet transit, an observer must either be exceedingly lucky, or observe a large number of stars. We will quantitatively assess these challenges in the next sections.

5.4.1 Detectability of Transit Dips

To assess the detectability of transits, let's consider the amount of light reaching an observer from a star before transit and during transit, as shown in Figure 5.7. Before transit, the amount of light reaching the observer is proportional to the cross-sectional area of the star, i.e.,

$$F_{\text{before transit}} \propto A_\star \qquad (5.17)$$

where F here refers to the flux of light reaching an observer.

During transit, part of the surface area is blocked by the planet. If we observe at visible wavelengths, it's a decent assumption that the planet contributes virtually no light of its own, since Wien's law (Equation (2.4)) tells us that the planet's blackbody emission peaks at much longer wavelengths.[3] Then,

$$F_{\text{during transit}} \propto A_\star - A_{\text{planet}}. \qquad (5.18)$$

---

[3] The situation does change at infrared wavelengths, a fact exoplanet scientists can exploit to help determine the planet's properties.

Thus, the change in the amount of light reaching the observer, $\Delta F = F_{\text{before transit}} - F_{\text{during transit}}$, is proportional to A_{planet}. The fractional reduction in light during the transit is then:

$$\frac{\Delta F}{F_{\text{before transit}}} = \frac{A_{\text{planet}}}{A_\star} \tag{5.19}$$

$$= \frac{R_{\text{planet}}^2}{R_\star^2}. \tag{5.20}$$

If a star's radius is known, a planet's radius can be deduced from a single transit observation using Equation (5.20). Alternatively, for a given theoretical planet, this equation provides the expected transit signal strength.

Let's see what this equation implies, practically speaking. Suppose we have a Jupiter-like planet, with $R_J = 7 \times 10^4$ km orbiting a Sun-like star with $R_\star = 7 \times 10^5$ km. Then:

$$\frac{\Delta F}{F} = \frac{R_{\text{planet}}^2}{R_\star^2} \tag{5.21}$$

$$= \left(\frac{7 \times 10^4 \text{ km}}{7 \times 10^5 \text{ km}}\right)^2 \tag{5.22}$$

$$= \frac{1}{100}. \tag{5.23}$$

In other words, the starlight will only dim by 1% if a Jupiter-like planet transits a Sun-like star.

To visualize what this means, consider Figure 5.8. When performing real observations, observations have associated noise, so to detect this small dip, the noise level must be less than the size of the dip. Supposing we have a dip of 1% (0.01), we may want a noise level of 0.005. In this case, the overall *signal-to-noise ratio (SNR)* of the observations must be 1/0.005 = 200. A SNR of 200 is high, but isn't unreasonable for good ground-based telescopes and, in fact, transits by Jupiter-like

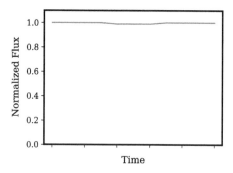

Figure 5.8. A theoretical transit lightcurve (normalized flux versus time) for a Jupiter-sized planet around a Sun-sized star, which causes only a 1% reduction in stellar flux.

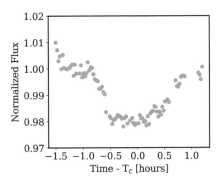

Figure 5.9. Transit lightcurve (normalized flux versus time from center of transit) of WASP-2 (showing the signature of its companion, WASP-2b), obtained at Vassar College's Class of 1951 Observatory by (then) undergraduate student Alexandra Trunnell. Notice the very different y-axis scale for this plot, as compared to Figure 5.8.

planets can be readily observed with small telescopes. For example, Figure 5.9 shows a lightcurve of a Jupiter-like planet obtained with Vassar College's Class of 1951 Observatory 32'' (0.8 m) diameter telescope.

However, let's consider the case of an Earth-like planet, which, for simplicity, we'll assume has a radius of 7×10^3 km. In this case,

$$\frac{\Delta F}{F} = \left(\frac{7 \times 10^3 \text{ km}}{7 \times 10^5 \text{ km}} \right)^2 \tag{5.24}$$

$$= \frac{1}{10^4} \tag{5.25}$$

and the corresponding SNR would need to be about 2×10^4. This is exceedingly difficult precision for ground-based telescopes to achieve, since Earth's fluctuating atmosphere is an ever-present source of noise. This also implies that there is an *observational bias* toward detecting large planets. In other words, since large planets are easier to detect, the absence of detected small planets should not be interpreted as a lack of small planets.

5.4.2 Transit Probability

As discussed above, transits are only observed if the planet and star are aligned from the observer's point of view—otherwise, the planet will not appear to pass in front of the star. Figure 5.10 shows the maximum inclination a planetary orbit can take and still transit the star, and also demonstrates how this angle, $\theta_{\max}$, decreases with increasing planetary semimajor axis.

Because the planet has a finite size, $\theta_{\max}$ also depends on the planet's size. If the planet is larger, it can have a slightly higher inclination and still have part of its disk transit in front of the star. More precisely, as we can see in Figure 5.10

$$\sin \theta_{\max} = \frac{R_\star + R_p}{a_p} \tag{5.26}$$

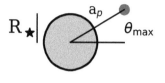

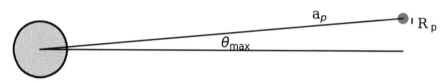

Figure 5.10. Visualization of the maximum inclination angle that allows a transit to be seen by an observer located on the right-hand side of the figure. Note that the size of the angle depends on the planet's semimajor axis.

where $R_\star$ and R_p are the radii of the star and planet and a_p is the semimajor axis of the planet's orbit. For small angles, $\sin \theta \approx \theta$, and this simplifies to

$$\theta_{max} \approx \frac{R_\star + R_p}{a_p}. \tag{5.27}$$

The probability of a transit occurring can then be computed by dividing θ_{max} by the range of possible inclinations the orbit can take. Since inclinations can randomly be anywhere between 0° and 90° (0 to $\pi/2$), we divide by $\pi/2$.

$$P_{transit} \approx \frac{\theta_{max}}{\pi/2} \tag{5.28}$$

$$\approx \frac{2(R_\star + R_p)}{\pi a_p} \tag{5.29}$$

$$\approx \frac{R_\star + R_p}{a_p} \tag{5.30}$$

(where we have assumed π is not too different from 2). Note that as the planet's semimajor axis increases, the probability of transit decreases. This tells us that there is a strong observational bias toward discovering transiting planets with small semimajor axis, a_p.

Let's also plug in some numbers. For a Jupiter-like planet orbiting a Sun-like star, the probability that its orbit is aligned just right for an observer to see a transit is:

$$P_{transit} \approx \frac{R_\odot + R_J}{a_J} \tag{5.31}$$

$$\approx \frac{7 \times 10^5 \text{ km} + 7 \times 10^4 \text{ km}}{5 \times 1.5 \times 10^8 \text{ km}} \tag{5.32}$$

$$\approx 0.001. \tag{5.33}$$

That means an observer would need to observe 1000 stars to reasonably hope to see a transit... and that's not even taking into account Jupiter's 12 year orbital period.

5.4.3 Determining Semi-major Axis from Transit Observations

While the depth of a transit lightcurve provides the radius of the planet (or, more precisely, the ratio of the planet's radius to the star's radius), the transit timing (duration and time between transits) can provide the planet's semimajor axis.

The time between two transits is simply the planet's orbital period. The transit duration—the time between the start and end of the transit—is given by:

$$T_{tr} = \frac{\text{distance traveled during transit}}{\text{velocity during transit}}. \tag{5.34}$$

As can be derived from Figure 5.11; this is equal to:

$$T_{tr} = \frac{2R_\star + 2R_p}{v_{\text{orb}}}. \tag{5.35}$$

For a circular orbit, the orbital velocity is given by the distance traveled divided by the period, or $v_{\text{orb}} = \frac{2\pi a_p}{P_p}$, where a_p and P_p refer to the semimajor axis and period of the planet's orbit, respectively. Substituting, we find:

$$T_{tr} = \frac{(2R_\star + 2R_p)P_p}{2\pi a_p} \tag{5.36}$$

or

$$\frac{T_{tr}}{P_p} = \frac{(2R_\star + 2R_p)}{2\pi a_p}. \tag{5.37}$$

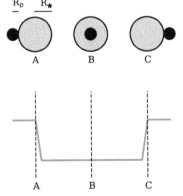

Figure 5.11. Relationship between configuration of star and planet and lightcurve features.

Therefore, if one measures the transit duration (T_{tr}), the time between transits (P_p), and R_p (from the transit depth), the planet's semimajor axis, a_p can be determined with Equation (5.37). An example of deriving R_p, P_p, and a_p from a transit lightcurve is given as a homework exercise.

5.4.4 Space Telescopes and Transit Detections

Large transiting exoplanets orbiting bright stars create lightcurve dips that are easily observable from even small ground-based telescopes. In fact, the first transiting exoplanet (HD 209458b) was identified concurrently by researchers using a 0.8 m telescope (Henry et al. 2000) and a 0.1 m telescope (Charbonneau et al. 2000). However, this planet was not *discovered* with this technique. In fact, it was discovered with the radial velocity technique, and, following its discovery, photometric monitoring allowed for the detection of its transit signature. Because of the low transit probability for any given star, a large number of stars must be monitored to detect planets with the transit technique. For this reason, for many years, the transit technique lagged far behind the radial velocity technique in detecting exoplanets.

This all changed with the launch of the Kepler Space Telescope in 2009 (Borucki et al. 2011). The Kepler telescope was specifically designed for exoplanet transit observations. To maximize the probability of detecting exoplanets, it was designed with a large field of view (meaning, the size of the patch of sky seen in a single exposure) that could monitor over 100,000 stars at a time in one part of the sky. In addition, its position in space means that it's not subject to the fluctuations in brightness caused by the Earth's atmosphere. Thus, the SNR of the data could be very high, and the Kepler Space Telescope was able to detect even Earth-sized planets. Thanks to the Kepler Space Telescope, thousands of exoplanets have been discovered with the transit technique, including the smallest planets detected to date (Borucki et al. 2011).

In 2018, NASA launched the Transiting Exoplanet Survey Satellite (TESS; Ricker et al. 2015). This satellite monitors 200,000 of the brightest stars near the Earth to look for transit signals. In contrast to Kepler, this satellite therefore looks around the entire sky, rather than staring at a single patch. In addition, since TESS targets bright stars, stars with detected planets can be more easily followed up with additional observations.

5.5 Gravitational Microlensing

Planets can also be detected with a technique made possible by the principles of general relativity. According to general relativity, since gravity curves space, light can actually bend around very massive objects like stars. If one star travels behind another star, from the perspective of an observer, the light from the background ("source") star will bend around the foreground star, creating multiple images of the source star for the observer (see Figure 5.12). Since the foreground star bends the light via gravity, we call it a *gravitational lens* star. If the observer's telescope had sufficient angular resolution, it would be able to image these multiple lensed images. However, the set of lensed images is too small to distinguish with current facilities.

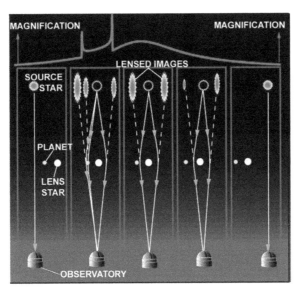

Figure 5.12. A cartoon demonstrating how gravitational microlensing can be used to detect planets. A lightcurve, shown above, is produced as a foreground lens star and planet pass in front of a background source star, from the perspective of an observer. The broad peak is produced by the lens star, while the additional bumps are produced by the lens planet. Image source: David Bennett and Sun Hong Rhie.

Instead, the multiple lensed images blend together in the observer's image, and make the source star appear as if it has temporarily brightened. In such a case, when the lensed images appear blended together, the technique is called gravitational *micro*-lensing. The observer can therefore use a lightcurve to look for this temporary brightening, which would indicate the passage of a background source star behind a foreground lens star.

How is this effect used to detect planets? If a planet is orbiting the lens star, its gravity will also act as a lens, splitting and amplifying the light from the source star, and causing its own bump(s) in the lightcurve. As seen in Figure 5.12, these bumps will be smaller than the main bump, and occur at a different time depending on where the planet is in its orbit, and how far away it is from its parent star. Typically, the large bump due to the star lasts about a few months, while the bumps due to the planet last a few days.

Since deriving the equations underlying this effect is beyond the level of this text, we'll simply state some of the results. First, by measuring the size and location of the planet's lightcurve bump, this technique provides the planet mass and star-planet separation (which can be converted into a lower limit on the planet's semimajor axis with some additional information about the lens star). In order to detect the planet, it must be far enough from the lens star for its gravity to be distinguished from the lens star's gravity, but close enough to the lens star that its lightcurve bump overlaps the lens star's bump. Thus, in contrast to the radial velocity and transit techniques, gravitational microlensing is sensitive to planets with moderate (not too small, but not too large) semimajor axes. Also, since the lensing event can dramatically brighten the source star, these events can be detected even for planets at very large distances from the Earth. One disadvantage of this technique is that it requires the observation of chance alignments of two stars, meaning that microlensing effects can only be detected by large surveys monitoring many stars over a long period of time. In addition, since the chance alignment of source and lens stars only occurs once, and the planet is often very far from Earth, the planet is unlikely to ever be observed a second time.

5.6 Direct Imaging

Direct imaging is perhaps the most conceptually simple means of detecting exoplanets, but remains one of the most technically challenging. As discussed in Section 5.1 and demonstrated in Figure 5.1, the primary challenge to direct imaging of exoplanets is the large star-planet brightness contrast combined with small star-planet separations. The angular separation (observed separation on the sky) of a star and planet is given by

$$\theta[\text{arcsec}] = \frac{a_p}{d} \frac{360}{2\pi} \times 3600 \tag{5.38}$$

where a_p is the semimajor axis of the planet, d is the distance to the star from Earth, and we have measured the angle in units of arcsec—the standard angular unit used by astronomers (recall the discussion in Section 5.3). Let's plug in some numbers to see how feasible this technique might be. Suppose we were to observe the nearest star to Earth, Alpha Centauri, at a distance of 4×10^{16} m (van Leeuwen 2007). Recalling that current ground-based observatories can create images as sharp as $\theta \sim 0.02$ arcsec in size, we could theoretically detect planets as close as 0.026 au to this star! So, at first glance, this is a feasible technique for nearby stars, although the minimum a_p we can detect increases linearly with the star's distance from Earth.

However, our calculation accounts only for the separation of the star and planet, and does not take into account the brightness contrast. An image of a star is never infinitely small, but is spread out, due to two effects. First, the telescope itself spreads out the light due to an effect called *diffraction*. It turns out that the larger the telescope, the sharper (less spread out) the image gets. The second effect, known as *atmospheric seeing*, is due to the rays of light from the star being refracted by the different layers of Earth's atmosphere. The blurring due to atmospheric seeing can vary depending on the telescope's location, and the nightly weather. Because the star's image is spread out, the planet image appears in the "wings" of the star's image, and the more extreme the brightness contrast of the star and planet, the harder it can be to locate the planet within the wings of the stellar image.

Space telescopes deal with these challenges by orbiting above Earth's atmosphere, avoiding atmospheric seeing altogether. However, space telescopes are smaller than the largest ground-based telescopes, due to the costs of launching large objects into space. Ground-based telescopes can be built much larger, minimizing the effects of diffraction, but must therefore correct for the effect of the Earth's atmosphere. They utilize a technique known as *adaptive optics*, in which one of the telescope mirrors is constantly deformed to counteract the blurring of the Earth's atmosphere. For both

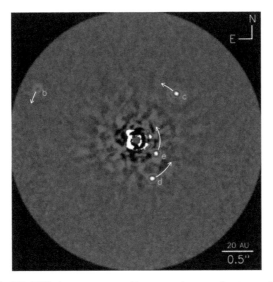

Figure 5.13. Image of the HR 8799 planetary system. By convention, exoplanets are given the name of the star plus a letter, beginning with "b" (for example, HR 8799 b). The designations are given in the order of detection, which is why the outermost planet (which was easiest to detect) has the designation "b", and the innermost is "e". Image Source: NRC-HIA/C. Marois/W/M. Keck Observatory.

ground- and space-based telescopes additional techniques can be used to remove stellar light, including *coronagraphy*, where the light from the star is physically blocked (just as you might use your hand to block the light from the Sun when trying to see an airplane), and image subtraction (where an image of the star is subtracted from the image of the star plus planet). With the use of these techniques, several planets have been directly imaged. One such system of planets, around the star HR 8799, and detected with ground-based telescopes, is shown in Figure 5.13 (Marois et al. 2008). The dark and light blobs near the center of the image show the residual noise that remains after attempting to remove the host star's light from the image, while the orange labeled globes are the true planets. As we would predict for this technique, the planets detected are bright (and, therefore, large, and young; see Section 6.3.1) and far from their parent star.

5.7 Properties of Known Exoplanets

5.7.1 Observational Biases in Exoplanet Observations

Before discussing the properties of discovered extrasolar planets, it's important to revisit the observational biases inherent in their detection. Recall that an observational bias means that some types of planets are easier to detect than others. When that's the case, the true distributions of planet properties must be considered with those biases in mind. As an example, imagine a universe in which there are equal amounts of Jupiter-mass and Earth-mass planets. If the Jupiter-mass planets are above the detection limit of current instruments, but Earth-mass planets are below the detection limit, we would only be able to detect Jupiter-mass planets. Suppose after a year of observing we detect

100 Jupiter-mass planets and 0 Earth-mass planets. Is it correct to say that the universe has only Jupiter-mass planets, and no Earth-mass planets (outside of the solar system)? Of course not. This may simply be a reflection of our observational bias. In some cases, the observational biases can be quantified and corrected for. For example, suppose a technique is 10× more likely to detect Jupiter-mass planets compared to Earth-mass planets, and scientists find 100 Jupiter-mass planets and 10 Earth-mass planets. Knowing that the detection technique is 10× more sensitive to Jupiter-mass planets, the scientists would conclude that the *actual* number of Earth-mass planets is likely to be about the same as that of Jupiter-mass planets.

In the case of the *radial velocity technique*, planets that are *massive* and *close* to their parent star (i.e., with small a_p) are easiest to detect. As of 2020, the most sensitive instruments can detect planets that cause a 1 m s^{-1} velocity shift (e.g., Mayor et al. 2003) sufficient to detect a 3 Earth-mass planet at 0.05 au orbiting a Sun-like star, or a ~Neptune-mass planet at 1 au orbiting a Sun-like star. The *transit technique* most readily finds planets that are *large* and *close* to their parent star. However, with its high sensitivity, the Kepler Space Telescope has detected planets as small as Mercury (Barclay et al. 2013). *Gravitational microlensing* finds planets that are *massive*, and at *moderate distance* from their parent stars. The *direct imaging technique* most readily detects planets that are *bright* (which depends both on temperature and size) and *far* from their parent star, making it complementary to the other available detection techniques. The sensitivity to high brightnesses also has the interesting effect of making *young* planets easier to detect. As we'll learn later in Chapter 6, young planets are brighter because they have residual energy from their formation.

The complex set of observational biases also affects where discovered planets are located in our Galaxy. Figure 5.14 shows the distances (from Earth) to host stars of

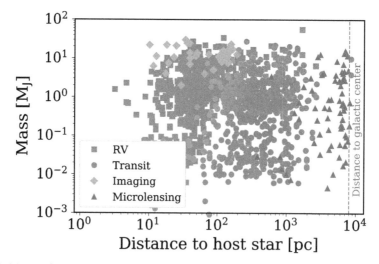

Figure 5.14. Masses of detected extrasolar planets versus their distance from Earth. For this and future plots, note that the mass is not directly determined from the data. For the radial velocity method, the mass shown is the minimum mass, which assumes an edge-on orbital inclination. For the transit method, the mass is inferred from the planet's radius. Distances are given in a standard unit of distance used in astronomy, call a parsec (pc for short), equal to ~3 × 10^{16} m. Data from the NASA Exoplanet Archive (NASA 2019).

detected extrasolar planets. Small distances make stars appear brighter, improving the SNR of transit and RV observations. In addition, closer distances imply larger angular separations (refer back to Equation (5.38)), making direct imaging observations easier. However, since the Kepler Space Telescope has detected most transiting exoplanets, transit detections are clustered around a "sweet spot" for Kepler, representing stars in its field of view that were not too faint, and not too bright, for accurate lightcurves. TESS, which will observe the brightest nearby stars, will detect many planets closer to Earth. Finally, the gravitational microlensing technique tends to find the farthest planets from Earth. This technique can easily detect stars as far as the galactic center, due to the large brightness changes that occur during the microlensing event. However, observers must survey regions of the Galaxy with a high density of stars to increase the likelihood of observing the chance alignment of foreground lens and background source stars.

5.7.2 Planet Masses

A histogram of observed (minimum) masses of discovered exoplanets is shown in Figure 5.15. A few features are worth noting. First, the histogram shows a strong increase in population toward lower masses. Second, there are a large number of planets with masses higher than any seen in our solar system (where, of course, the maximum planet mass is 1 M_J—one Jupiter-mass). We need to consider these facts in light of observational biases. In fact, since small mass planets are harder to detect with all detection techniques, the observational bias in all cases would cause us to detect *fewer* small planets. Therefore, there must truly be more smaller planets than larger planets. Since high mass planets are easier to detect, we can't say whether the prevalence of very high mass planets (greater than a Jupiter-mass) means our solar system is unusual. However, their existence alone tells us that Jupiter is not the most massive possible planet that can be formed.

5.7.3 Mass and Semimajor Axis

Figure 5.16 shows a plot of planet mass versus star-planet separation. Points are color-coded according to their detection technique. Before we discuss the interesting

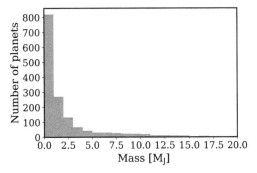

Figure 5.15. Histogram of masses for confirmed exoplanets, with data from the NASA Exoplanet Archive (NASA 2019).

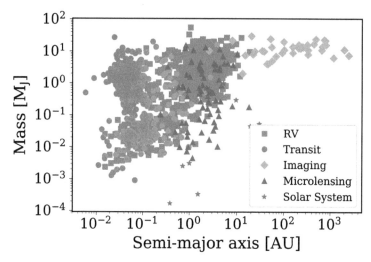

Figure 5.16. Mass versus semimajor axis for confirmed exoplanets, with data from the NASA Exoplanet Archive (NASA 2019). Purple stars show solar system planets.

details in this plot, it's important to discuss the axes. In order to put all of these points on the same y-axis, radial velocity planets must be plotted according to their minimum mass, while transit-detected and directly imaged planets must have their radii and brightnesses, respectively, converted to mass. Thus, there are some uncertainties in the y-axis due to these inferences. In order to put all of these points on the same x-axis, radial velocity and transit-detected planets are plotted according to their semimajor axes, while directly imaged and microlensing-detected planets are plotted according to the *observed* star-planet separation. This observed separation is not necessarily the size of the orbit, since the planet may have an elliptical orbit and because the orbit of the planet can be inclined with respect to the plane of the sky. As an extreme example, consider a very large orbit that is viewed nearly edge-on, and observed when the planet is right next to the star from our perspective. Its apparent separation would in that case be very small, even though its orbit was very large. Therefore, these caveats should be kept in mind when considering this plot in detail.

Nevertheless, we can notice many interesting features in this plot. First, it shows the observational biases of each technique. Both the radial velocity and transit technique tend to detect planets close to their star, but the transit technique is more heavily biased toward small separations. The radial velocity technique cannot find very small planets, but the transit technique can find even Mercury-sized planets. Gravitational microlensing finds planets at moderate distances (1–10 au) from their parent stars. The most distant planets are found by direct imaging, and these planets tend to be large as well. This plot also highlights the explosion in the detection of planets that was brought about by the Kepler space telescope. Finally, we can see the observational bias creating a lack of planets at small masses, with the least massive planets detected increasing with separation. Close to the star, even small planets can be found; far from the star, only the most massive planets can be found.

Once we've noted the observational biases, can we say anything about the planet populations themselves? Because close-in planets are easiest to detect, it's interesting to note that few planets are found within 0.01 au. There appears to be some region close to the star within which planets cannot form and/or survive. In addition, we can see that close to the star, small planets are much more commonly observed, even though they are actually harder to detect than large planets. So it seems as if, at least in the ~0.01–0.2 au region or so, small planets are much more common than large planets. We can also see that many directly-imaged planets have large separations from their stars. Since the semimajor axis of Pluto's orbit is 39 au, we can tell that some planets are able to form at (or be transported to) much larger separations than we have seen in the solar system.

Another surprising result of this plot is the very large number of planets near 0.01 Jupiter masses. The mass of the Earth is about 0.003 Jupiter masses, while the next most massive planet in the solar system, Uranus, has a mass of about 0.05 Jupiter masses, leaving a noticeable lack of planets in the few–10 Earth mass range. Exoplanetary systems, however, seem to have no trouble forming planets in this mass range. These planets have come to be called *super Earths*.

A final surprising result in this plot is the large number of Jupiter-mass (and larger!) planets that are found very close to their parent stars. Recall that in the solar system, the closest planet to the Sun— Mercury—has a semimajor axis of 0.4 au. In contrast, there are exoplanetary systems in which Jupiter-like planets reside as close-in as ~0.02 au. As we'll discuss in later sections, it may be difficult to *form* planets this close to a star, so the planets may have formed elsewhere and traveled to their current positions. These planets, for obvious reasons, have come to be known as *hot Jupiters*.

5.7.4 Orbital Eccentricities

Figure 5.17 shows orbital eccentricity plotted against orbital semimajor axis. These particular points are from radial velocity measurements, which most readily provide orbital eccentricity. As we know, the radial velocity technique most easily detects planets in smaller orbits. The observational bias regarding eccentricity is not obvious, but it has been determined that due to uncertainties in the data,

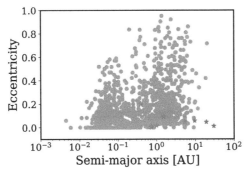

Figure 5.17. Plot of eccentricity versus semimajor axis for confirmed exoplanets detected with the radial velocity technique (blue circles), with data from the NASA Exoplanet Archive (NASA 2019). Red stars show the solar system planets.

eccentricities tend to be slightly overestimated when planets are in near circular orbits (Zakamska et al. 2011). In other words, there are more nearly-circular orbits than it would appear from this plot alone.

In spite of this small caveat, two incredibly interesting results can be derived from this plot. First, recall that orbital eccentricities for planets in the solar system are very small. The orbital eccentricity of Mercury (and dwarf planet Pluto) is near 0.2. For all other solar system planets, orbital eccentricities are below 0.1. Although there are certainly many exoplanets with small eccentricities, there are also many with eccentricities greatly exceeding solar system values. Therefore, the planet formation process must be able to produce planetary systems with uniformly low eccentricities, as well as systems with large eccentricities.

Another interesting feature of this plot is that eccentricities appear to decrease as separations decrease. In particular, planets with separations below ∼0.02 au are all in nearly-circular orbits. This is the result of something called tidal circularization, which is similar to the tidal synchronization discussed in Section 4.3.2. Imagine a system in which the planet's rotational period equals its orbital period, and then consider what happens when the planet is near perihelion or near aphelion. Similar to the case of a mismatch between orbital and rotation speeds, the non-circular orbit will cause the tidal bulge to be misaligned with the star-planet line. This mismatch will cause a torque that, over time, will cause the orbit to circularize. Since tidal force is strongly dependent on the separation between the interacting bodies (recall Equation (4.34)), the circularization will happen most rapidly for close-in planets.

5.7.5 Mass–Radius Relationships, or Planet Density

Since density is mass divided by volume ($\rho_{\text{avg}} = M/(\frac{4}{3}\pi R^3)$) for a sphere, both the planet's mass and radius must be known to determine its density. Therefore, planet density is typically only determined for planets that have both transit and radial velocity data. Note that if transit data exist for the planet, then the inclination is known to be nearly zero (otherwise the planet's orbit would not pass in front of the star). Therefore, the planet's minimum mass is essentially equal to its actual mass. Figure 5.18 shows radii and masses of exoplanets whose densities have been measured. Solid curves show expected relationships for planets with different bulk compositions (Howard et al. 2013). Perhaps not surprisingly, models predict that planets made of denser materials, like iron, are smaller than planets made of less dense gases, like hydrogen. However, it's important to realize that the density of a material depends on its pressure, so there is no simple scaling of planetary size with composition. In addition, we can see an interesting turnover in the curves at high mass. While we might expect planets to continually get larger as they increased in mass, it turns out that at some point, they become so massive that the force of gravity begins to overwhelm the addition of mass, and the planet will actually decrease in size as it increases in mass.

Let's now look at the planetary data. We can see, first, that smaller planets are more likely to be solid-rich, and more massive planets are more likely to be gas-rich—a prediction we might have made based on our knowledge of the solar system.

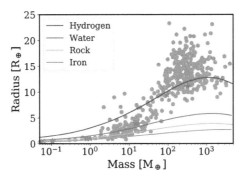

Figure 5.18. Plots of planet radius (in Earth radii) versus planet mass (in Earth masses), adapted from Howard et al. (2013), using data from the NASA Exoplanet Archive (NASA 2019). Solid curves show expected mass–radius relationships for planets with different bulk compositions (Howard et al. 2013). Purple stars show solar system planets.

Therefore, it seems that there may be a limit on how massive a planet can be and still remain solid-rich. However, there is a surprising discrepancy between the models and observed radii for Jupiter-like planets, with many Jupiter-like planets having radii even larger than those expected for planets made of pure Hydrogen. This discrepancy is likely due to a dynamical "puffing-up" of exoplanet atmospheres when the planets are particularly close to their parent stars, although the detailed physical causes are still under debate.

We can also see that low-mass extrasolar planets display a wide range of compositions, including planets that are very dense, those that are rocky like the Earth, and those that are more water-rich, and therefore more similar to our ice giants Uranus and Neptune. In general, low-mass exoplanets seem to show a wide array of different compositions—perhaps even compositions that are not represented in our solar system.

5.7.6 The Habitable Zone

The detection of extrasolar planets throughout the Galaxy has renewed interest in the search for life beyond Earth. One concept that helps guide this search is the idea of a *habitable zone*—the region around a star where temperatures are "just right" for life to exist (for fun, this is also sometimes referred to as the "Goldilocks zone"). What defines "just right"? Well, at least on Earth, liquid water is required for life, so a habitable zone might be defined as the region around a star within which water can exist on the planet's surface.

However, there are many complications that arise when considering where water might exist on an extrasolar planet. To calculate a planet's expected temperature, we could use its equilibrium temperature (recall Equation (2.15)), but first we'd need to know the planet's albedo. In addition, as we explored in Homework Question 2 in Chapter 2, a planet's true temperature is not necessarily its equilibrium temperature—it can be affected by the presence of an atmosphere, or by additional energy sources. Finally, the phase of water is determined not only by temperature, but also by pressure. A planetary body (like Jupiter's moon Europa) might not have liquid water on its

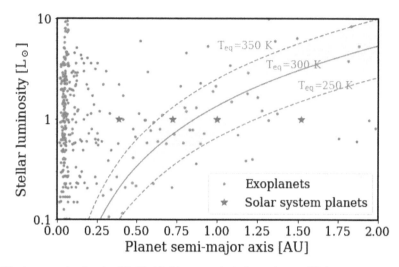

Figure 5.19. An example of a simplified habitable zone calculation, using equilibrium temperatures between 250 K and 350 K, computed with albedos of 0. On top of the curves, we show the locations and stellar luminosities of known exoplanets (orange circles; NASA 2019) and the solar system planets (purple stars).

surface, but rather deep inside its interior. Of course, this assumes liquid water is necessary and sufficient for life—are either of those true? We'll have to leave that to the biologists to answer. In spite of these complications, astronomers find the concept of a habitable zone useful for quantifying the likelihood of finding life elsewhere in the universe, and for fine-tuning searches for habitable worlds; they continue to discuss, debate, and research the details.

In Figure 5.19 we show an example of a simplified habitable zone calculation. Here, we ignore any sophisticated details, and use the equilibrium temperature equation to define regions around a star where temperatures are between 250 and 350 K (i.e., hovering near the 273 K melting temperature for Earth's surface, but allowing for some wiggle room). Notice how the habitable zone region depends on the stellar luminosity, moving outwards as the star brightens. In addition, the width of the habitable zone also depends on stellar luminosity. Finally, notice the interesting fact that, with this definition, we might (very incorrectly) assume Venus is habitable! This highlights the idea that a habitable zone is a useful concept, but doesn't tell the whole story.

5.7.7 Concluding Thoughts: How Can You Stay Up-to-date?

With the first exoplanet around a Sun-like star being detected less than a few decades ago (Mayor & Queloz 1995), our knowledge of exoplanets continues to grow at incredible speed. Therefore, this chapter is far from being the final word on exoplanets. An interested reader can easily update their understanding of the properties of known exoplanets by exploring the data themselves. Up-to-date data can be downloaded directly from the NASA Exoplanet Archive: https://exoplanetarchive.ipac.caltech.edu (NASA 2019) and be used to re-make many of the plots in this section.

5.8 Important Terms

- star/planet contrast;
- Doppler shift;
- radial velocity;
- minimum mass;
- transit/transit lightcurve;
- signal-to-noise ratio (SNR);
- observational bias;
- gravitational lens/gravitational microlensing;
- diffraction;
- atmospheric seeing;
- adaptive optics;
- coronagraphy;
- super Earth;
- hot Jupiter;
- the habitable zone.

5.9 Chapter 5 Homework Questions

1. **The size of radial velocity signals:**[4]

   (a) The maximum radial velocity of a star with an orbiting exoplanet is given by Equation (5.8) for an edge-on orbit. Use this equation to estimate the radial velocity signature an alien would see if they tried to detect Jupiter orbiting the Sun at 5 au.

   (b) Estimate the maximum radial velocity of a Sun-like star with an Earth-mass companion at 1 au, in an edge-on orbit.

   (c) The first radial velocity measurements (Mayor & Queloz 1995) could detect stellar velocities of about $10\,\mathrm{m\ s^{-1}}$. Would they have been sensitive enough to detect a Jupiter-like planet? An Earth-like planet?

   (d) A modern spectrograph, like the High Accuracy Radial Velocity Planet Searcher (HARPS; Mayor et al. 2003) can detect stellar velocities of about $1\,\mathrm{m\ s^{-1}}$. Is it sensitive enough to detect a Jupiter-like planet? An Earth-like planet?

   (e) Figure 5.5 shows a portion of the solar spectrum that includes the sodium D lines–one of which is centered at 5 889.950 Angstroms. Given the radial velocity you calculated above for a Jupiter-like planet, and the Doppler shift equation (Equation (5.1)), what would be the expected wavelength shift for this line? What would be the expected wavelength shift for an Earth-like planet? Evaluate the size of your answer by comparing to the sizes of physical objects.

---

[4] Adapted from John Asher Johnson.

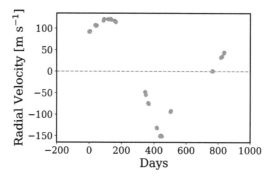

Figure 5.20. Radial velocity of HD 111232 measured by the HARPS instrument (data from the "HARPS RV Bank"—Trifonov et al. 2020).

(f) The resolution of a spectrograph is measured by its spectral resolution, $R = \frac{\lambda}{\Delta\lambda}$, where $\Delta\lambda$ is the smallest wavelength shift that can be measured, and λ is the wavelength observed. For the HARPS instrument, which has a spectral resolution of $R = 120{,}000$, what is $\Delta\lambda$ when observing near the 5 889.950 Å Sodium D line?

(g) Discuss how the minimum wavelength shift detectable by HARPS compares with the expected wavelength shift for a Jupiter-like planet, and for an Earth-like planet. Based on what you find, discuss why exoplanet researchers may need to measure the shifts of a large number of lines, rather than just one.

2. **Working with real radial velocity data:[4]**

(a) Use the equation for maximum radial velocity signal for an edge-on orbit (Equation (5.8)) along with a generalized version of Kepler's third law (Equation (3.49)), to find an expression for the mass of a planet that depends on $M_\star$, v, P_p and any necessary constants (but not a_p). Check your answer by making sure the expression you found will produce an answer that has units of mass.

(b) Figure 5.20 shows radial velocity measurements of HD 111232 measured by the HARPS (Mayor et al. 2003) instrument. Using this plot, and the equation you just derived, estimate the period, semi-major axis, and minimum mass of this star's companion—HD 111232 b. The mass of HD 111232 is 0.78 $M_\odot$ (Mayor et al. 2004).

(c) Given your answers to the last question, what type of planet do you think this might be? Is it similar to any particular planet(s) in the solar system? Is it different from solar system planets? Explain your answer.

(d) Notice that the positive and negative peaks have different absolute values. What kind of orbit would produce this asymmetry? Draw a diagram to explain your answer.

3. **Working with real transit data:**

(a) Figure 5.9 shows a transit lightcurve for the star WASP-2. Using this curve, and a radius for WASP-2 of 0.843 $R_\odot$ (Daemgen et al. 2009) estimate the radius of its planetary companion, WASP-2 b. Express your answer in units of Jupiter radii (R_J).

(b) Use the equation for transit duration, $\frac{T_{tr}}{P_p} = \frac{R_\star + R_p}{\pi a_p}$, and the generalized form of Kepler's third law (see Question 2), to derive an expression for a_p in terms of $R_\star$, R_p, and the transit duration, T_{tr} (but not P_p).

(c) Using Figure 5.9 and the expression you just derived, estimate the semimajor axis of WASP-2 b's orbit. The mass of WASP-2 is 0.89 $M_\odot$ (Daemgen et al. 2009).

(d) Given the radius and semimajor axis you found, what type of planet do you think this might be? Is it similar to any particular planet(s) in the solar system? Is it different from solar system planets? Explain your answer.

References

Barclay, T., Rowe, J. F., Lissauer, J. J., et al. 2013, Natur, 494, 452

Borucki, W. J., Koch, D. G., Basri, G., et al. 2011, ApJ, 736, 19

Brault, J., & Neckel, H. 1987, Spectral Atlas of Solar Absolute Disk-Averaged and Disk-Center Intensity, ftp://ftp.hs.uni-hamburg.de/pub/outgoing/FTS-Atlas

Charbonneau, D., Brown, T. M., Latham, D. W., et al. 2000, ApJL, 529, L45

Daemgen, S., Hormuth, F., Brandner, W., et al. 2009, A&A, 498, 567

Gaia Collaboration, Prusti, T., de Bruijne, J. H. J., et al. 2016, A&A, 595, A1

Gaia Collaboration, Brown, A. G. A., & Vallenari, A. 2018, A&A, 616, A1

Henry, G. W., Marcy, G. W., Butler, R. P., et al. 2000, ApJL, 529, L41

Howard, A. W., Sanchis-Ojeda, R., Marcy, G. W., et al. 2013, Natur, 503, 381

IPAC/NASA Exoplanet Science Institute 2019, NASA Exoplanet Archive, https://exoplanetarchive.ipac.caltech.edu

Marois, C., Macintosh, B., Barman, T., et al. 2008, Sci, 322, 1348

Mayor, M., Pepe, F., Queloz, D., et al. 2003, Msngr, 114, 20

Mayor, M., & Queloz, D. 1995, Natur, 378, 355

Mayor, M., Udry, S., Naef, D., et al. 2004, A&A, 415, 391

NASA/Goddard Space Flight Center 2020, Space Science Data Coordinated Archive: Planetary Fact Sheet, https://nssdc.gsfc.nasa.gov/planetary/factsheet/

Ricker, G. R., Winn, J. N., Vanderspek, R., et al. 2015, JATIS, 1, 014003

Trifonov, T., Tal-Or, L., Zechmeister, M., et al. 2020, A&A, 636, A74

van Leeuwen, F. 2007, A&A, 474, 653

Zakamska, N. L., Pan, M., & Ford, E. B. 2011, MNRAS, 410, 1895

Introductory Notes on Planetary Science
The solar system, exoplanets and planet formation
Colette Salyk and Kevin Lewis

Chapter 6

Planetary Interiors

In this chapter, we'll discuss how we know what we know about planetary interiors. In particular, we'll discuss the measurement of bulk planet density and moment of inertia (which tells us whether or not the planet has a core). In addition, we'll try to understand *why* some planets are differentiated and some are not by discussing the sources of energy in planetary interiors—the gravitational energy of formation and energy from radioactive decay.

6.1 Bulk Density

Even knowing the *average* (also called *bulk*) *density* of a planet can tell us something interesting about its interior. A planet with a large fraction of iron, for example, should have a higher bulk density, while a planet with a large fraction of ice should have a relatively lower bulk density. In this section, we'll discuss the basic techniques for measuring the bulk density of planetary bodies.

6.1.1 Measuring a Bulk Density

The bulk density of a planet is approximately given by

$$\rho = \frac{M}{\frac{4}{3}\pi R^3} \tag{6.1}$$

where we have assumed here that the planet is well-approximated by a sphere of radius R, and M is the planet's mass. Thus, the bulk density of a planet can be determined if we can measure the radius and mass of the planet.

If a planetary body can be directly imaged and spatially resolved (meaning that its shape can actually be seen in the image), then the *angular size* of the body, θ, can be directly measured. The angular size is the apparent size of the object on the sky as shown in Figure 6.1, and depends on both the actual size of the object and its distance as:

doi:10.1088/2514-3433/abb198ch6

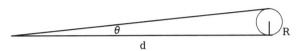

Figure 6.1. Figure demonstrating the relationship between angular size (θ), true size (R for radius), and distance (d). For actual bodies in the solar system, as viewed from Earth, R would need to be much smaller relative to d for this figure to be to scale, and θ would correspondingly be much smaller as well.

$$\theta = \frac{2R}{d}, \tag{6.2}$$

where R is the radius of the planet and d is the distance between the planet and the observer. (This should look familiar, as the very similar Equation (5.38) shows the angular separation of a star and planet at a distance d.)

Measuring angular size for close/large objects is relatively straightforward. Measuring distance is somewhat trickier, although distances of planets relative to the Earth–Sun distance were determined hundreds of years ago using geometry. These distances also rely on angles on the sky. The angle between a planet's position when it appears farthest from the Sun provides a measure of its orbital size relative to Earth's orbital size, as shown in Figure 6.2. While the relative sizes of orbits have been known for centuries, the absolute Earth–Sun distance (i.e., the value of 1 au) was finally pinned down via measurements of the time it takes radio signals to bounce off of the planets Venus and Mars.

The mass of a planet is most easily determined if it has one or more moons. If we think back to the derivation of Kepler's third law (Equation (3.49)), we'll see that such a law would also apply to the moons of a planet, with P and a representing the Moon's period and semimajor axis, respectively, and $M_\odot$ replaced by the planet's mass. Thus, by measuring the size and period of a moon's orbit, we can determine the planet's mass. This does assume that we know the value of G, however, and G was not measured until 1798 (with the so-called Cavendish experiment).

If a planetary body has no moons, measuring its mass becomes more difficult. The mass of such bodies can be determined by analyzing their interactions with other bodies in the solar system. Alternatively, if they're visited by a human-made satellite, the human-made satellite's orbit can be tracked to determine the mass of the planetary body.

Measuring the mass of the Earth is somewhat different from measuring the densities of other planets. For one, before the start of space exploration, the whole Earth could not be seen at once. However, the radius of Earth has been known since at least 200 BC, when Eratosthenes used the angles of the Sun's shadow at two points on Earth to infer its size. The period of the Moon's orbit around the Earth is straightforward to measure, and the Moon's distance can now be determined with incredible accuracy by measuring the travel time of laser light from the Earth reflected off of mirrors left on the Moon's surface by Apollo astronauts and other lunar missions. As with other planets with moons, the Moon's orbital distance and period can then be used to measure Earth's mass using Kepler's third law.

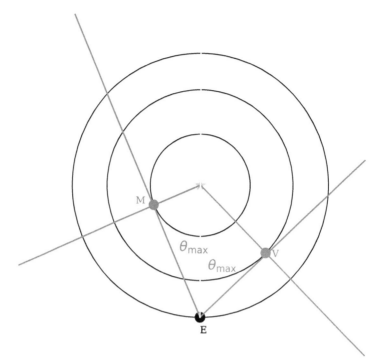

Figure 6.2. The relative distances to the planets can be determined with geometry (and were determined by Copernicus hundreds of years ago). This figure shows how the maximum angular separations of interior planets (Mercury and Venus) from the Sun (called "maximum elongation", and labeled θ_{max}) are dependent on their distance from the Sun. The maximum elongation is determined by observing the planet and Sun in the sky, and repeatedly determining the angle between them. The line connecting the Earth and the planet's point at maximum elongation is tangent to the planet's orbit, and thus forms the right triangles shown. Using simple geometry, the planetary semimajor axis is then given by 1 au × $\sin(\theta_{max})$. (Note: This figure assumes the orbits are circular and co-planar; in reality, the maximum elongation varies due to orbital inclinations and eccentricities.)

6.1.2 Interpretation of Bulk Density

Table 1.1 gives the bulk densities of the solar system planets, plus the Moon and the dwarf planet Pluto. Some care must be taken when comparing bulk densities to densities of everyday materials, such as rock, ice, metals, and gases, since the compression of such materials at high pressure in the interiors of planets increases their densities. So, for example, even if a planet were made entirely of rock, it would have a bulk density higher than the 3000 kg m^{-3} density of a typical rock on Earth's surface. Nevertheless, it's clear that the gas and ice giants have much lower bulk densities than the terrestrial planets, and that the density of Pluto (and most other small bodies in the outer solar system) lies somewhere in-between. It bears noting, however, that the bulk density of the gas giants is only ~5 times lower than that of the terrestrial planets, while, for example, air at atmospheric pressure on the surface of the Earth is more than 1000 times less dense than a rock. Therefore, if these planets are composed primarily of gas, the interiors of the giant planets must be at much higher pressures than we are accustomed to on the surface of Earth.

The bulk densities of the solid bodies also tell an interesting story. Venus and Earth have similar masses and densities, and so their bulk compositions are likely similar. In fact, more detailed analysis shows that they are composed of both dense, iron-rich material, and less dense rocky material. Mercury is much less massive than Venus and Earth, and so its interior should be much less compressed, and yet it has a density similar to Earth. This suggests that Mercury has a higher proportion of iron-rich material. For the Moon and Mars, it's not immediately obvious whether their lower bulk density (relative to Earth) is due to decreased compression, or less iron-rich material, but more detailed studies show that they both have less iron-rich material than the Earth (much less, in the case of the Moon). On the other hand, Pluto's density is much lower than even "uncompressed" rocks. Its location far from the heat of the Sun allowed it to incorporate a higher fraction of lower density ices into its solid body.

6.2 Moment of Inertia and Interior Structure

Beyond determining the bulk density of a planetary body, it can be useful to know its interior structure. Is the interior of uniform composition (is the body *undifferentiated*)? Or does the planet perhaps have a high density core (is it *differentiated*)? These questions can be answered by measuring a planet's *moment of inertia*.

6.2.1 Review of Moment of Inertia and the Moment of Inertia Factor

A body's moment of inertia can be thought of as a rotational analog of mass. Just as linear momentum $\vec{p} = m\vec{v}$, we can write $\vec{L} = I\vec{\omega}$, where $\vec{L}$ is the angular momentum and $\vec{\omega}$ is the angular speed (the rotational analogs to $\vec{p}$ and $\vec{v}$, respectively), and I is moment of inertia. For discrete bodies with mass m_i spinning about a rotation axis at perpendicular distances r_i, the moment of inertia of the system is given by

$$I = \Sigma_i r_i^2 m_i. \tag{6.3}$$

However, for a solid body, the moment of inertia is given by

$$I = \int_0^M r^2 \, dm, \tag{6.4}$$

where r is the perpendicular distance of a small amount of mass, dm, to the rotation axis.

This definition tells us that, when calculating how easy it is to cause angular acceleration, we must consider not only the mass of the body, but also how that mass is distributed within the body. The more centrally condensed the mass is, the smaller I is, and the easier it is to cause an angular acceleration. As an example, if we compute I for an infinitely thin spherical shell, we find that $I = \frac{2}{3}MR^2$, where M is the mass of the shell and R is its radius. For a constant density sphere, we would find a lower value: $I = \frac{2}{5}MR^2$. For real planetary objects, we're unlikely to have mass concentrated in a hollow shell, but quite likely to have the mass centrally condensed, in a core. Therefore, for planetary bodies, it's usually the case that $I < \frac{2}{5}MR^2$. For the Earth, for example, $I \approx 0.33MR^2$ (Williams 1994).

Since all of the calculations for moment of inertia for a sphere yield an expression of the form $I = \alpha MR^2$, the parameter α is given its own name—the *moment of inertia factor*. (Inconveniently, α is the same symbol as is used for angular acceleration, so be aware of the context when you see this symbol.) If we rewrite this expression we see that

$$\alpha = \frac{I}{MR^2}. \qquad (6.5)$$

If we could measure this moment of inertia factor for a planet, we could determine whether it has a constant density interior or a centrally condensed interior. In other words, we could determine whether it is undifferentiated or differentiated. And, the smaller the value of α, the more centrally condensed the planet is. In the next sections, we'll discuss how we can actually go about measuring α for a planet.

6.2.2 Equatorial Bulges and the Flattening Parameter

Just as a spun clump of pizza dough flattens out, spinning planetary bodies tend to bulge at their equators, and planets are shaped like oblate spheroids, rather than perfect spheres. In this section, we'll develop some intuition about the magnitude of that bulging effect. Given an equatorial radius a and a polar radius c, we'll estimate the *flattening parameter*

$$f = \frac{a - c}{a} \qquad (6.6)$$

for a constant density oblate spheroid.

First, let's review the concept of a *potential*. Just as we have gravitational force, $\vec{F}_G = -\frac{GMm}{r^2}\hat{r}$ and an associated potential energy, $U_G = -\frac{GMm}{r}$, where $F_G = -\frac{dU_G}{dr}$, we can define a gravitational potential, $V_G = -\frac{GM}{r}$, which is the potential energy per unit mass. Since the planet is rotating, the planet experiences a centrifugal force, which has magnitude $F_C = \frac{mv^2}{r} = m\omega^2 r$, where r is the distance to the rotation axis and ω is the angular speed. The associated potential energy is therefore $U_C = -\frac{m\omega^2 r^2}{2}$, and the associated potential is $V_C = -\frac{\omega^2 r^2}{2}$.

Particles (of a given mass) anywhere on the surface of a planet must have the same potential energy—otherwise, they would tend to redistribute themselves until they did. Therefore, we say that the surface is an *equipotential surface*. If we imagine a point on a planet's pole, point p, and a point on its equator, point e, the potentials at those two points must be equal:

$$V_{G,p} + V_{C,p} = V_{G,e} + V_{C,e}. \qquad (6.7)$$

Since the distance between the pole and the rotation axis is 0, $\nabla_{C,p} = 0$. Therefore, we have:

$$\nabla_{G,p} = \nabla_{G,e} + \nabla_{C,e} \tag{6.8}$$

$$-\frac{GM}{c} = -\frac{GM}{a} - \frac{\omega^2 a^2}{2}. \tag{6.9}$$

Let's define the magnitude of the bulge to be $\Delta r = a - c$, i.e., the difference in size between the equatorial radius and the polar radius. Then, we can rewrite the gravitational potential at the equator as follows:

$$-\frac{GM}{a} = -\frac{GM}{c + \Delta r} \tag{6.10}$$

$$= -\frac{GM}{c\left(1 + \dfrac{\Delta r}{c}\right)}. \tag{6.11}$$

And, we can simplify this expression by using a Taylor expansion, which tells us that $\frac{1}{1+\frac{\Delta r}{c}} \approx 1 - \frac{\Delta r}{c}$ since $\frac{\Delta r}{c} \ll 1$.

$$-\frac{GM}{a} \approx -\frac{GM}{c}\left(1 - \frac{\Delta r}{c}\right) \tag{6.12}$$

$$\approx -\frac{GM}{c} + \frac{GM\Delta r}{c^2}. \tag{6.13}$$

With this approximation, Equation (6.9) can be rewritten as:

$$-\frac{GM}{c} \approx -\frac{GM}{c} + \frac{GM\Delta r}{c^2} - \frac{\omega^2 a^2}{2} \tag{6.14}$$

$$\frac{GM\Delta r}{c^2} \approx \frac{\omega^2 a^2}{2}. \tag{6.15}$$

The flattening, f, is then given by:

$$f \approx \frac{\Delta r}{c} \approx \frac{1}{2}\frac{\omega^2 a^2 c}{GM} \approx \frac{1}{2}\frac{\omega^2 a^3}{GM}, \tag{6.16}$$

where in the last step, we again use the fact that Δr is very small to say that $c \approx a$.

Our "derivation" above is not completely correct, because of our approximations, and because we did not fully account for the three-dimensional shape of the planet. However, Newton did determine a more precise estimate of the flattening, which differs by our estimate by only a constant factor, so we've captured the essential intuition. Although it is still only an approximation, Newton's more precise equation is:

$$f = \frac{5}{4} \frac{\omega^2 a^3}{GM}. \tag{6.17}$$

The expression should match our intuition. The equatorial bulge will be largest if the planet is large, and rotating quickly. However, bulging will be reduced if the planet is very massive, as the gravitational force of the large mass will tend to keep the body compressed.

6.2.3 Relationship Between Moment of Inertia and Flattening

When deriving an estimate of the flattening parameter, we assumed that the planetary body had a constant density, but this is unlikely to be the case. In fact, the interior structure of the planet affects the degree of flattening caused by its rotation. Intuitively, we would expect that an empty thin shell would experience the maximum amount of flattening, since it has little centralized mass to resist the centrifugal force. On the other hand, a very centrally condensed body would have the minimum amount of flattening, since its central core would strongly resist the centrifugal force. Since moment of inertia is a measure of the internal structure, the flattening depends on the moment of inertia.

Remember that the moment of inertia factor for a sphere of mass M and radius R, $\alpha = \frac{I}{MR^2}$. Recognizing that $R \approx a$ (the equatorial radius), we will assume here that

$$\alpha \approx \frac{I}{Ma^2}. \tag{6.18}$$

Therefore, the flattening parameter for an arbitrary interior structure might have similarities to Equation (6.17), but should also depend on α, such that larger α corresponds with greater bulging. The derivation of this expression is beyond the scope of this text, so we simply state here the expression for f as a function of α known as the Darwin–Radau approximation (e.g., de Pater & Lissauer 2010):

$$f = \frac{\frac{5}{2} \frac{\omega^2 a^3}{GM}}{1 + \frac{25}{4}\left(1 - \frac{3}{2}\alpha\right)^2}. \tag{6.19}$$

To provide us with some intuition about this expression, we can substitute in a few values for α and see if the value of f matches our expectations. For a constant density sphere ($\alpha = 2/5$), we recover Newton's equation, $f = \frac{5}{4}\frac{\omega^2 a^3}{GM}$, as expected. For a thin spherical shell ($\alpha = 2/3$), we should find a larger f, and indeed, we would find $f = \frac{5}{2}\frac{\omega^2 a^3}{GM}$—two times larger than the flattening in the constant density case. For a centrally condensed body ($\alpha < 2/5$), we should find a smaller f. Letting $\alpha = 1/3$, for example, we find $f = \frac{40}{41}\frac{\omega^2 a^3}{GM}$.

Now that we have this expression, we can use measurements of a planet's degree of flattening to infer the value of its moment of inertia factor, assuming we know its size, mass, and rotation rate. As a homework exercise, you'll use images of the asteroid Ceres to figure out whether or not it is a differentiated body.

6.3 Energy in Planetary Interiors

6.3.1 Energy of Formation

If we drop an object on the surface of the Earth, it tends to move toward a state of lower potential energy. In the process, since total energy is conserved, the lost potential energy (*mgh*, where *h* is the height which the object falls) is converted to kinetic energy. Similarly, on a larger scale, a particle in space in the vicinity of a planet would tend to be attracted to that planet. In the process, it would lose potential energy and gain kinetic energy.

In fact, the process of building an entire planet can be thought of as a series of such events, in which particles are brought from far distances to the surface of the growing planet, one by one converting their initial potential energies into kinetic energy. Once the planet forms, the kinetic energy becomes the motion of all of the atoms and molecules inside of the planet. We call this remaining energy the planet's *energy of formation*.

Using this idea, we can derive an expression for the energy of formation. Figure 6.3 illustrates the setup for this derivation. Particles of mass *dm* start at infinite distance from the planet and are brought to the surface of the growing planet, which has an instantaneous mass *m* and radius *r*. Remembering that the gravitational potential energy of a mass *m* is given by $U_G = -\frac{GMm}{d}$ (see Equation (3.54)),[1] where M is the other body interacting with m via the gravitational force and

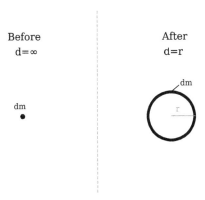

Figure 6.3. Illustration of the procedure for computing the energy of formation of a planetary body. Particles of mass *dm* start at infinite distance and are brought to the surface of the growing planet, with instantaneous mass *m* and radius *r*. We assume the mass gets magically spread around the surface of the planet, like a spherical veneer.

---

[1] We use *d* in place of the usual *r* for this equation, as we need to reserve the *r* symbol to mean the radius of the growing planet.

d is the distance between the two masses, the initial potential energy of the particle of mass dm is

$$U_i = -\frac{G\,m\,dm}{\infty} \tag{6.20}$$

$$\approx 0. \tag{6.21}$$

The final potential energy of the particle is:

$$U_f = -\frac{G\,m\,dm}{r}. \tag{6.22}$$

Therefore, the energy difference, or the energy converted to the planet's internal energy, is given by $U_i - U_f = \frac{Gmdm}{r}$.

While building the planet, we can imagine that the mass gets converted into a spherical veneer on the planet, of radius r and thickness dr. If we assume a constant density, ρ, then the instantaneous mass of the planet is given by $m = \frac{4}{3}\pi r^3\rho$, while the mass of the thin veneer can be expressed as $\rho\,4\pi r^2 dr$. The total energy of formation is then calculated by integrating over dr from 0 to R, where R is the final radius of the planet.

$$E_{\text{form}} = \int \frac{G\,m\,dm}{r} \tag{6.23}$$

$$= \int_0^R \frac{G\left(\frac{4}{3}\pi r^3\rho\right)(\rho\,4\pi r^2 dr)}{r} \tag{6.24}$$

$$= G\frac{16}{3}\pi^2\rho^2 \int_0^R r^4\,dr \tag{6.25}$$

$$= G\frac{16}{3}\pi^2\rho^2\frac{1}{5}R^5. \tag{6.26}$$

To find a final expression in terms of the mass and radius of the planet, we'll use the replacement:

$$\rho^2 = \left(\frac{M}{\frac{4}{3}\pi R^3}\right)^2. \tag{6.27}$$

Then,

$$E_{\text{form}} = G\frac{16}{3}\pi^2\frac{M^2}{\left(\frac{4}{3}\right)^2\pi^2 R^6}\frac{1}{5}R^5 \tag{6.28}$$

$$= \frac{3}{5} \frac{GM^2}{R}. \tag{6.29}$$

To get some intuition about the value of this quantity, let's compute it for Earth. The energy of formation for the Earth is

$$E_{\text{form},\oplus} = \frac{3}{5} \frac{GM_\oplus^2}{R_\oplus} \tag{6.30}$$

$$= \frac{3}{5} \frac{(6.67 \times 10^{-11} \text{m}^3 \text{ kg}^{-1} \text{ s}^{-2})(6 \times 10^{24} \text{ kg})^2}{6 \times 10^6 \text{ m}} \tag{6.31}$$

$$\approx 2 \times 10^{32} \text{ J}. \tag{6.32}$$

To put this into context, the amount of energy Earth receives from the Sun in one year is $\sim 10^{24}$ J—significantly less than the energy of formation. The energy associated with Earth's orbit is -10^{33} J. The energy of formation, therefore, is not quite at the same magnitude as orbital energies, but it's not that far off either.

Although this is the standard expression for energy of formation, it can be helpful to know whether more massive planets would tend to have higher or lower formation energies—a point that is not immediately obvious from this expression, since it depends on both M and R. If we assume all planets have the same density, then $R \propto M^{1/3}$. Therefore,

$$E_{\text{form}} \propto \frac{M^2}{M^{1/3}} \tag{6.33}$$

$$E_{\text{form}} \propto M^{5/3}. \tag{6.34}$$

More massive planets have a larger energy of formation.

6.3.2 Energy from Radioactive Decay

Another source of internal energy for planets is radioactive decay, in which a parent nucleus spontaneously changes to a daughter nucleus, releasing subatomic particles and energy. Of particular importance in the early solar system is the decay of $^{26}_{13}$ Al to $^{26}_{12}$ Mg. It's still an unresolved question why ^{26}Al was present in the early solar system, as it is produced both by massive stars (which could have injected it into the early solar system via winds or supernovae), and by the breakup of nuclei by energetic cosmic rays that permeate space. However, based on chemical measurements from meteorites, it appears to have been widespread in the early asteroid belt.

In the ^{26}Al decay process, a proton (p) in the Al atom is converted to a neutron (n), positron (e^+) and neutrino (ν_e), and this conversion releases $\Delta E = 3.12$ MeV $\approx 5 \times 10^{-13}$ J of energy (Castillo-Rogez et al. 2009), i.e.,:

$$p \rightarrow n + e^+ + \nu_e + \Delta E. \tag{6.35}$$

Although it's not possible to predict when a single atom might decay, the decay rate for a collection of atoms can be characterized by a quantity known as the half-life ($\tau_{1/2}$), which is the time it takes for half of the parent element to decay to the daughter element:

$$\% \text{ parent remaining} = \left(\frac{1}{2}\right)^{\frac{t}{\tau_{1/2}}} \tag{6.36}$$

where t is the elapsed time. For ^{26}Al, the half-life is 7.16×10^5 yr (Samworth et al. 1972).

^{26}Al was present in the early solar system, but the percent remaining after the full solar system lifetime of 4.5 Gyr is $\left(\frac{1}{2}\right)^{\frac{4.5\times10^9}{7.16\times10^5}} = \left(\frac{1}{2}\right)^{6285}$. Since this number is exceedingly small, there is essentially no primordial^{26}Al remaining in the solar system today, and ^{26}Al is referred to as an *extinct radionuclide*.

Given that there is no ^{26}Al remaining today, we can estimate how much energy was provided to planets by the decay process. To calculate the total number of decay reactions, we divide the total mass of ^{26}Al in a planetary body by the mass of a single ^{26}Al atom, as follows:

$$\# \text{ of reactions} = \frac{\text{Total } ^{26}\text{Al mass in body}}{\text{Mass of single } ^{26}\text{Al atom}}. \tag{6.37}$$

We can rewrite the numerator in terms of the mass of the planetary body, M, if we also know the mass percent of Al and the original mass ratio of ^{26}Al to the dominant (stable) Al isotope, ^{27}Al:

$$\text{Total } ^{26}\text{Al mass in body} = M \times \% \text{ Al} \times \left(\frac{^{26}\text{Al}}{^{27}\text{Al}}\right). \tag{6.38}$$

For the Earth, for example, $M = 6.0 \times 10^{24}$ kg, mass % Al is 0.014, and the original mass ratio of the two isotopes measured in meteoritic materials is 5×10^{-5} (MacPherson et al. 1994 and references therein). Therefore, the total mass of ^{26}Al that existed in the early Earth would be $\approx 4 \times 10^{18}$ kg. Given that the mass of a single ^{26}Al atom is about 26 atomic mass units, or about 4.3×10^{-26} kg, we find that there would have been $\approx 10^{44}$ decay reactions. The total energy released by this set of reactions is then (# of reactions) $\times \Delta E = (10^{44}) \times (5 \times 10^{-13} \text{ J}) \approx 5 \times 10^{31}$ J.

We can compare this to some other relevant energies to give this number context. For example as we'll see in Chapter 7, the energy associated with the dinosaur extinction impact event is of order 10^{24} J, so radioactive decay provides much more total energy than such an event (although, of course, the impact would release the energy much more quickly). We calculated the energy of formation for Earth to be about 2×10^{32} J, so the radioactive decay is somewhat less important—but only by about a factor of 4.

However, let's see how the energy of radioactive heating compares with the energy of formation for smaller bodies. As can be seen in Equation (6.38), the energy released due to radioactive decay is proportional to the planet mass, $E_{\text{decay}} \propto M$. On the other hand, as shown in Equation (6.34), $E_{\text{form}} \propto M^{5/3}$. Therefore, the energy of formation will drop more dramatically for smaller masses than will the energy for radioactive decay. For example, for a planet that is 10 times less massive than the Earth, the energy from decay of ^{26}Al becomes $(1/10) \times 5 \times 10^{31}$ J $= 5 \times 10^{30}$ J. The energy of formation becomes $(1/10)^{5/3} \times 2 \times 10^{32}$ J $= 4 \times 10^{30}$ J. Therefore, the energy from radioactive decay is larger than that of formation for this small body. For some small bodies in the solar system, therefore, radioactive decay can provide the dominant source of internal energy. However, it's important to remember that the half-life of ^{26}Al is only 7.16×10^5 yr. As we'll discuss in Chapter 9, the formation process takes up to a few Myr. Therefore, if a planetary body is to gain a significant amount of energy from the decay of ^{26}Al, it must form quickly.

6.3.3 Melting and the Formation of Planetary Cores

We know that the Earth has a core, based on its moment of inertia, along with evidence from seismology and many other observations. How and why did its core initially form? If planets get hot enough to melt, denser material like iron naturally settles toward the planet's center, leaving lighter material above. This process of differentiation forms the core, and, above it, the mantle. Now that we know how much energy the planet gains during its formation, we can ask whether this is sufficient to melt the planet and cause differentiation.

In order for a planet to melt, it would first need sufficient energy to reach its melting temperature, and then would need sufficient energy to change phase. The amount of energy required to reach the melting temperature depends on the mass of the body, M and its specific heat, C, as well as the temperature change required to reach the melting point, ΔT:

$$E_{\Delta T} = MC\Delta T. \tag{6.39}$$

For rocks, a typical melting temperature is around 1500 K, while values for specific heat are approximately 1000 J kg^{-1}K^{-1}. For an Earth mass planet to reach melting temperature starting from an ambient 300 K, therefore, a total energy of $(6 \times 10^{24}$ kg$)(1000$ J kg^{-1}K$^{-1})(1200$ K$) \approx 7 \times 10^{30}$ J is required. The energy of formation for the Earth is about 2×10^{32} J, so a body the size of Earth can readily receive enough energy during formation to reach the melting temperature.

The amount of energy required to melt depends on the mass of the body, M, and its heat of fusion, ΔH_{fus},

$$E_{\text{melt}} = M\Delta H_{\text{fus}}. \tag{6.40}$$

For rocks, typical heats of fusion are around 3×10^5 J kg^{-1}. For an Earth-sized body, therefore, the energy required to melt is $(6 \times 10^{24}$ kg$) \times (3 \times 10^5$ J kg$^{-1}) \approx 2 \times 10^{30}$ J. This is less energy than is required to reach the melting point, so is unlikely to serve as a barrier to melting. The Earth, therefore, had enough energy of formation to melt and therefore differentiate.

But what about smaller planets? Looking at Equation (6.39) and knowing that the energy of formation scales as $M^{5/3}$, we can see that the temperature change experienced by a planet during its formation would be

$$\Delta T = \frac{E_{\text{form}}}{MC}, \tag{6.41}$$

which will scale as $\frac{M^{5/3}}{M}$ or $M^{2/3}$. In other words, smaller planets will not get heated to such high temperatures. We can also ask if there is a mass for which the gravitational energy is just enough to raise the planetary temperature to the melting point. This would occur when:

$$MC\Delta T = E_{\text{form}} \tag{6.42}$$

$$= \frac{3}{5}\frac{GM^2}{R}. \tag{6.43}$$

If we know the planet's average density, ρ, we can substitute for R to find:

$$MC\Delta T = \frac{3}{5}\frac{GM^2}{\left(\frac{3M}{4\pi\rho}\right)^{1/3}} \tag{6.44}$$

$$M = \left(\frac{C\Delta T}{\frac{3}{5}G\left(\frac{4}{3}\pi\rho\right)^{1/3}}\right)^{3/2}. \tag{6.45}$$

Substituting in $\Delta T = 1200$ K and $C = 1000$ J kg^{-1} K^{-1} as above, as well as $\rho = 3000$ kg m^{-3}, we find that this planet would have a mass of about 5×10^{22} kg—slightly smaller than Earth's Moon, and about the mass of Jupiter's moon Europa. Thus, large moons in the solar system have enough mass to be entirely melted due to the formation process. Smaller bodies do not have enough energy of formation to completely melt, but they may or may not differentiate, depending on whether they form quickly enough to take advantage of the heat from ^{26}Al and (and any other short-lived radionuclides).

6.3.4 Cooling Rates and Retention of Energy

When discussing planetary surfaces in Chapter 7, we will see that large terrestrial planets tend to be geologically active, while small bodies are not (unless they have some source of energy, such as tidal interactions). Now we know that more massive planets start with a larger energy of formation. But do they also retain that energy for longer than do smaller bodies? To answer that question, we have to consider not just the energy of formation, but also the cooling rate. If we know the cooling rate, Γ, then the characteristic timescale over which a planet loses its internal energy will be given by:

$$\tau_{\text{cool}} = \frac{E_{\text{form}}}{\Gamma}. \tag{6.46}$$

To derive an expression for Γ, we'll remind ourselves of the derivation of equilibrium temperature. In this derivation, we assumed the planet was in equilibrium, such that:

$$\text{Energy in per unit time} = \text{Energy out per unit time, eq} \qquad (6.47)$$

where "eq" here refers to equilibrium. We then noted that the energy out per unit time is given by the planet's luminosity, assuming the planet has a temperature T_{eq}. Thus, in equilibrium,

$$\text{Energy in per unit time} = \text{Energy out per unit time, eq} \qquad (6.48)$$

$$= 4\pi R^2 \sigma T_{eq}^4. \qquad (6.49)$$

If a planet is *not* in equilibrium, then the actual energy out per unit time will be given by the planet's luminosity, given its *true* temperature, T. In other words,

$$\text{Energy out per unit time} = 4\pi R^2 \sigma T^4. \qquad (6.50)$$

The cooling rate of the planet is then given by the difference between the energy *actually* going out and the energy going in:

$$\Gamma = 4\pi R^2 \sigma (T^4 - T_{eq}^4). \qquad (6.51)$$

We can see from this expression that when the planet reaches equilibrium, its cooling rate goes to zero. This is expected, as the planet has reached equilibrium and should not change temperature with time. In addition, we can see that the cooling rate has a strong dependence on temperature. Therefore, as the planet cools, its cooling *rate* will decrease. So we would expect that if a planet begins with a lot of internal energy, it will cool quickly. But as its temperature approaches the equilibrium temperature, its cooling rate will slow and the planet could take some time to reach equilibrium.

As one example, let's consider the planet Jupiter. Since Jupiter has no surface, we have to figure out an appropriate temperature to use to compare with its equilibrium temperature. We use a quantity known as effective temperature, which is defined to be $T_{eff} = \left(\frac{L}{\sigma 4\pi R^2}\right)^{1/4}$, which roughly corresponds with the temperature at the atmospheric level that is optically visible. Jupiter's effective temperature is about 124.4 K, while its equilibrium temperature is about 109.5 K. Therefore, Jupiter is still cooling. You will compute and investigate this cooling rate as a homework question.

Now that we know the cooling rate, we can ask whether this helps us understand why large planetary bodies are more likely to be geologically active. The cooling rate, $\Gamma \propto R^2$, so if all planets have roughly the same density, then $R \propto M^{1/3}$ and $\Gamma \propto M^{2/3}$. Since the energy of formation scales as $E_{form} \propto M^{5/3}$

$$\tau_{cool} = \frac{E_{form}}{\Gamma} \qquad (6.52)$$

$$\propto \frac{M^{5/3}}{M^{2/3}} \tag{6.53}$$

$$\propto M. \tag{6.54}$$

So indeed, the characteristic timescale over which a planet retains its heat of formation is larger for larger mass planets, and it's not surprising that larger planets are more likely to be geologically active 4.5 billion years after their formation.

Although our discussion here captures some of the basics of planetary cooling, the details are much more complex. While a full discussion of planetary cooling is beyond the scope of this introductory text, considering the following. Γ depends on the temperature of the planet's surface, but what actually determines the temperature of a planet's surface? It turns out that the surface temperature depends on a number of factors, including how the planet's heat is transported from its interior to its surface, and the geological properties of the surface (for example, whether the planet has plate tectonics). Therefore, the actual cooling rate of a planet can be quite complex, and seemingly similar planets (like the Earth and Venus) can end up with very different geologic histories.

6.4 Interiors of Solar System Planets

6.4.1 Interior Structure

Figure 6.4 summarizes our basic understanding of solar system planetary interiors. Remember that information about planetary interiors is all *inferred* from techniques like those discussed in this chapter, and not directly measured by, say, drilling down to the center of the planet. Additional techniques that we did not cover here include detailed measurements of the gravitational pull on human-made satellites, and—for the Earth, Moon, and Mars (where we have been able to place seismometers)—seismology.

All planets of the solar system appear to be differentiated. For the terrestrial planets, we know that the heat of formation was likely sufficient to melt the planet and cause core formation. Although we can determine the approximate size of cores of the terrestrial planets from bulk density and moment of inertia, it is more difficult to determine the presence of an inner core, due to the lack of a strong density contrast with the liquid outer core. Note that Mercury's core is much larger than the other terrestrial planets as a fraction of the planet's total size, possibly as a result of a large impact early in its history (Asphaug & Reufer 2014). Conversely, the Moon only has a comparatively small core, also as a result of the impact between the Earth and a Mars-size body that led to the formation of the Moon (Canup & Asphaug 2001). The InSight seismic mission at Mars should reveal more detail about that planet's interior structure in coming years.

For the giant planets, the story may be different, as we'll see in Chapter 9—their dense cores of heavier elements, including metals, silicates, and ices, surrounded by gaseous atmospheres, are likely the result of their formation mechanism. Since different formation mechanisms predict different gas giant interiors, measurements of bulk densities and moments of inertia are crucial for determining how planets formed both within and outside of the solar system.

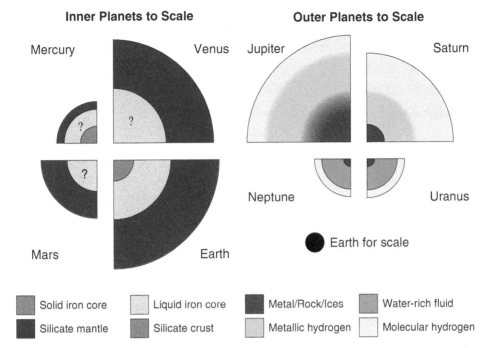

Inner Planets to Scale

Outer Planets to Scale

| Solid iron core | Liquid iron core | Metal/Rock/Ices | Water-rich fluid |
| Silicate mantle | Silicate crust | Metallic hydrogen | Molecular hydrogen |

Figure 6.4. Illustration of our current understanding of planetary interiors. Note that the interiors of the planets other than Earth shown here are models, and have not yet been fully determined by spacecraft observations in most cases. Design based on a similar figure by N. Strobel (Strobel 2020).

6.4.2 Differentiation and Earth's Surface Composition

For the Earth, geologists also learn about Earth's interior, and the process of differentiation, by studying Earth's surface (crustal) composition. Figure 6.5 shows the composition (by number) of some of the most abundant elements in the Sun, ordered from most to least abundant. (Note that the abundances are provided on a log scale, such that, for example, hydrogen is nearly 10^4 times as abundant as oxygen, etc.) Since the Sun and planets ultimately formed from the same materials, the region in the vicinity of early Earth should have had roughly *solar composition*. As we can see in Figure 6.5 (and as we'll discuss in more detail in Section 9.6.2), some of the most primitive objects in the solar system—known as chondritic meteorites, or chondrites—do have a strikingly similar composition to the Sun. They are, however, depleted in very light elements like H, He, Ne, and Ar, which would have been in the gas phase in the early solar system, and were not readily incorporated into meteoritic parent bodies. Figure 6.5 shows that the Earth's surface also has a similar composition to the Sun, although with somewhat less of other light elements (including, for example, N and C).[2] All of these light, so-called *volatile elements* are colored blue to indicate that they are often found in the gas phase, and unlikely to form a large part of Earth's solid composition.

---

[2] Although highly depleted relative to solar composition, N and C are major components of Earth's atmosphere, and Earth's life-forms, respectively.

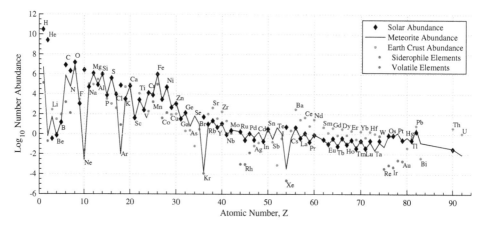

Figure 6.5. Elemental abundance of the Sun, primitive meteorites (called chondrites), and the Earth's crust (relative to Silicon). Data from Lodders et al. (2009), Rumble (2020).

A major difference between the Earth's surface composition and chondritic meteorite compositions arises, however, because the Earth has differentiated. During the segregation of Earth's core, *siderophile* (literally: iron-loving) elements, including Fe, Ni, and the valuable Platinum-group metals separate from other materials, leaving the Earth's rocky portion depleted in these elements. These siderophile elements are shaded red in Figure 6.5. They appear depleted in Earth's crust (and mantle), but are believed to be abundant in Earth's metal-rich core. The remaining elements include the *lithophile* (rock-loving) elements, which appear in relative abundance in Earth's crust and mantle.

6.5 Important Terms

- bulk/average density;
- angular size;
- undifferentiated/differentiated;
- moment of inertia;
- moment of inertia factor;
- flattening parameter;
- potential/equipotential surface;
- Darwin–Radau approximation;
- energy of formation;
- extinct radionuclide;
- solar composition;
- volatile;
- siderophile/lithophile.

6.6 Chapter 6 Homework Questions

1. **Average density:**

   In this question, we'll explore what a measurement of average density (also known as bulk density) can tell us about the interior of a planet.

   (a) Let's assume all planets are made of two components, of density ρ_1 and ρ_2, and masses M_1 and M_2, respectively. (For example, a planet may be made of a higher density metal core, plus a lower density rock mantle. Or, a comet might be made of rock plus lower density ice.) If the planet has a total mass M, and an average density ρ_{avg}, derive an expression for the mass of component 1 (M_1) as a function of M, ρ_{avg}, ρ_1 and ρ_2. Hint: Begin with the definition of density, and replace variables until everything is in terms of M, ρ_{avg}, ρ_1 and ρ_2.

   (b) Run the following tests of your expression to make sure it's correct. (If it's not, go back and find your error.) To run these tests, you will want to "plug" some hypothetical numbers into your expression:

      i. Let $\rho_2 = \rho_{avg}$. What do you expect M_1 to be in this case? Check that your expression gives this result.

      ii. Let $\rho_1 = \rho_{avg}$. What do you expect M_1 to be in this case? Check that your expression gives this result.

      iii. Let ρ_{avg} be something in-between ρ_1 and ρ_2. What do you expect M_1 to be in this case (roughly speaking)? If you test two different values of ρ_{avg} (but keep ρ_1 and ρ_2 fixed) how do you expect M_1 to change? Check that your expression gives the expected results.

   (c) Assuming component 1 represents the core of the planet, find an expression for the radius of the core, if it has mass M_1 and density ρ_1. (Note: You do not need the answers to parts (a) or (b) to answer this question.)

   (d) Assume that all planets are composed of an iron core, with a density of $\rho_1 = 8000$ kg m^{-3} and a rocky mantle, with $\rho_2 = 3000$ kg m^{-3}. Using the expressions you derived, the masses of the planets, the radii of the planets, and the average density of the planets, fill in the table below.

   (e) Discuss your findings. Which planets have the largest $\frac{R_{core}}{R_{pl}}$? Which have the smallest? Are there any trends to the data? What could cause the differences you see?

   (f) The estimated radii of the cores of each body are shown in the final column. How do these radii compare to the radii you derived? Explain how you would change your model to make it more physically realistic. In particular, what property of iron and rock might you want to know in addition to what is provided here (their uncompressed densities)?

Planet Name	Average Dens [kg m^{-3}]	Core Mass Frac (M_1/M)	Core Mass [kg]	Calculated R_{core} [km]	R_{core}/R_{pl}	Actual R_{core} [km]
Mercury	5427					~1600
Venus	5243					~3200
Earth	5514					3485
Mars	3933					~1700
Moon	3340					300–400

Note. Average densities from NASA's Planetary Fact Sheet (NASA 2020).

2. **Moment of inertia and differentiation:**
   In this question, we'll explore how actual measurements are used to measure a moment of inertia factor.

   (a) Rearrange Equation (6.19) to find an expression for α (recall that $\alpha = \dfrac{I}{Ma^2}$; see Equation (6.18)) as a function of the flattening, f, and other relevant parameters.

   (b) Calculate α for Ceres—a large asteroid—using the expression you just derived and the measured values of its equatorial and polar radii: 487.3 and 454.7 km, respectively (Thomas et al. 2005). In addition, you should assume that the period of rotation is 9.075 hours and the mass of the asteroid is 9.395×10^{20} kg.

   (c) Compare your value of α to the moment of inertia factors of an undifferentiated body (0.4) and that of the Earth (0.330 7; Williams 1994). What does your value of α suggest about the interior of the asteroid Ceres?

   (d) How do you think these parameters for Ceres were determined? Describe how you might measure a and c, the asteroid's total mass, and its rotation period.

   (e) Suppose Ceres had a rotation rate that was 10 times lower, but the same value of α. Compute the flattening parameter, f, and $a - c$, for this hypothetical body. Suppose the error bars on the measurement of the equatorial and polar radii are about 2 km in size (Thomas et al. 2005). Would it have been possible to detect the flattening in the case that Ceres' rotation rate were 10 times lower? Explain your answer.

   (f) Compute the rotation rate, ω, for Venus and compare to the rotation rate of Ceres. How many times slower does Venus rotate compare to Ceres? Based on this rotation rate and your answer to part (e), comment on the ease of measuring the moment of inertia factor for Venus.

(g) In Section 6.3.3; we found that if average density is constant, ΔT due to the formation of a planet scales as $M^{2/3}$. Use this expression, and the masses of Earth and Venus, to estimate the *ratio* of the temperature change for Earth and Venus.

(h) Given your calculation in part (g), and what you know about Earth's interior, would you infer that Venus is likely to be a differentiated body? Explain your answer.

3. **Cooling rates:**

In this question, we'll explore the cooling of Jupiter.

(a) Calculate the equilibrium temperature of Jupiter. Assume an albedo of 0.343 and an emissivity of 1. Use at least 2 significant digits for every term in your calculation (including the semimajor axis of Jupiter).

(b) Calculate the luminosity Jupiter would have if its actual temperature equaled its equilibrium temperature.

(c) For terrestrial planets, we can compare the actual surface temperature of the planet with the equilibrium temperature; for gas giants, we instead calculate something called the "effective temperature", T_{eff}, where the luminosity of the planet is given by $\sigma T_{\text{eff}}^4 4\pi R^2$. Jupiter's effective temperature is 124.4 K. Calculate its actual luminosity.

(d) Calculate the ratio of Jupiter's actual luminosity divided by the energy per unit time absorbed by Jupiter from the Sun. (Hint: You do not need to re-calculate the energy absorbed from the Sun.) What does this tell you about Jupiter?

(e) What is Jupiter's cooling rate (in W)? What is its cooling rate as a fraction of its total luminosity?

(f) Discuss what you learned, and speculate about what might be happening to Jupiter.

References

Asphaug, E., & Reufer, A. 2014, NatGe, 7, 564

Canup, R. M., & Asphaug, E. 2001, Natur, 412, 708

Castillo-Rogez, J., Johnson, T. V., Lee, M. H., et al. 2009, Icar, 204, 658

de Pater, I., & Lissauer, J. J. 2010, Planetary Sciences (Cambridge: Cambridge Univ. Press)

Lodders, K., Palme, H., & Gail, H.-P. 2009, Solar System, Landolt–Börnstein–Group VI Astronomy and Astrophysics, Vol. 4B (New York: Springer), 712

MacPherson, G. J., Davis, A. M., & Zinner, E. K. 1995, Metic, 30, 365

NASA Space Science Data Coordinated Archive 2020, Planetary Fact Sheet: https://nssdc.gsfc.nasa.gov/planetary/factsheet/

Samworth, E. A., Warburton, E. K., & Engelbertink, G. A. 1972, PhRvC, 5, 138

Strobel, N. 2020, Astronomy Notes, https://www.astronomynotes.com/solarsys/s8.htm, retrieved 2020 July 7

Rumble, J. R. (ed) 2020, CRC Handbook of Chemistry and Physics (101st ed.; Boca Raton, FL: CRC Press)

Thomas, P. C., Parker, J. W., McFadden, L. A., et al. 2005, Natur, 437, 224

Williams, J. G. 1994, AJ, 108, 711

Introductory Notes on Planetary Science
The solar system, exoplanets and planet formation
Colette Salyk and Kevin Lewis

Chapter 7

Planetary Surfaces

The study of planetary surfaces could easily encompass a text of its own (and the study of our own planet's surface can encompass a lifetime of study), so we must by necessity only scratch the surface of the topic of planetary surfaces. In this chapter, we'll focus on two of the most basic processes that shape planetary surfaces, namely, impact cratering and *tectonic activity*, which is a term that describes the large-scale stresses which act on the crust of a planet and affect its structure and surface expression. Tectonic forces give rise to many of the mountains and valleys we see on the surfaces of planets, and this topic connects naturally to Chapter 6, where we learned why some planets remain geologically active and some do not. We'll also discuss how impact craters can be used to study the history of the solar system, which we'll revisit in the chapter on planet formation. Finally, we'll end this chapter by briefly discussing the composition of planetary surfaces—including what determines their basic properties, and how we can study them with remote sensing.

7.1 Impact Cratering

7.1.1 Basic Physics of Cratering

One of the most universal geologic processes in the solar system is impact cratering. In fact, Jupiter's moon Io is the only solar system body for which impact craters have not been observed, due to rapid resurfacing from its many active volcanoes. Planetary surfaces are covered with impact craters because the solar system is full of small rocky and icy bodies, whose orbits can be perturbed over time by the larger planets. This can eventually lead to a collision with another planetary body, which will result in an impact crater on the larger object. How do the properties of an impactor relate to the properties of the crater it forms? Of course, we might guess that a large impactor will make a larger crater, and we'd be correct. However, if we drop marbles into sand, or watch a high velocity cratering experiment, we'll quickly see that the size of the crater created is larger than the size of the impactor, and sometimes much larger. The basic reason for this is that the impactor carries with it kinetic energy. When it impacts the surface, if the surface particles are free to move, the impactor will transfer some of its kinetic energy to the particles it collides with, causing them to be displaced from the surface. (The displaced particles are called *ejecta*.) With some very simple physical reasoning, we can derive a relationship between impactor energy and crater size.

For small impacts, the strength of the target material is the primary limit to crater formation, with much of the kinetic energy going into fracturing the target. The required energy will be roughly proportional to the volume of rock fractured (i.e., the size of the crater), which scales as R^3 for a crater of radius R. For larger craters (greater than a few kilometers diameter, on the Earth), this is surpassed by the gravitational potential energy required to form a crater, which we can roughly estimate. Figure 7.1 shows our setup for deriving the impactor energy-crater size relationship. Essentially, we'll imagine a crater as a hemisphere of radius R excavated into a planetary crust of density ρ_{crust}. This assumes a crater with similar depth and radius, which is roughly true for the initial "transient" crater (craters eventually evolve into different shapes, which are much shallower than they are wide). The creation of the crater requires lifting this volume of material from below the surface to above the surface—i.e., we have to lift it up a height R. How much energy would be required to lift the mass of the crater by a height R? This is simply given by the change in gravitational potential energy,

$$\Delta U_g = M_{crater} g R. \tag{7.1}$$

The mass of the material in the crater is given by $\frac{2}{3}\pi R^3 \rho_{crust}$, since we are assuming the crater is a hemisphere of radius R. Therefore,

$$\Delta U_g = \frac{2\pi}{3} R^3 \rho_{crust} g R \tag{7.2}$$

$$= \frac{2\pi}{3} \rho_{crust} g R^4. \tag{7.3}$$

The energy required to lift the material must be supplied by the impactor. In other words, the impactor's kinetic energy must equal

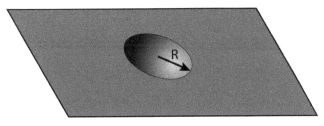

Figure 7.1. Setup for simple derivation of impactor energy-crater size relationship. We imagine the crater as being a hemisphere of radius R.

$$KE_{impactor} = \frac{2\pi}{3}\rho_{crust}gR^4. \tag{7.4}$$

Or, as a proportional relationship,

$$KE_{impactor} \propto R^4. \tag{7.5}$$

Empirical data and more sophisticated modeling tell us that this relationship is more complicated in detail, and give slightly different values for the constants and exponents in Equation (7.4). What are some of the details we're missing? As with small craters, some energy goes into fracturing, melting, and even vaporizing the previously intact crustal rocks. Further, we know that the impact ejecta are not simply lifted out of the crater; while much of the material from the transient crater falls back during formation of the final crater shape, some ejecta can be launched to large distances. This requires that the ejecta obtain the energy required to be lifted out of the crater *plus* the additional kinetic energy required to travel this distance.

Figure 7.2 (left) shows tektites—droplets of melted rock formed during an impact and commonly launched far beyond the crater itself. They bear a striking resemblance to raindrops, since they are formed as the molten rock travels through the Earth's atmosphere and simultaneously cools. Since it requires energy to melt and transport the rock, some of the impactor's energy must have been converted to those uses. Energy can also be used to change the mineral structure of the rocks; different crystal forms of the same chemical composition are called polymorphs. In Figure 7.2 (right), we see a phase diagram for SiO_2, which we often call quartz, but which can actually come in several different polymorphs (of which quartz is only one). In this figure, can see that coesite, as well as another SiO_2 polymorph called stishovite, are formed at high pressures and low temperatures—conditions typical of

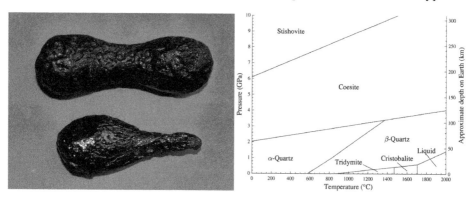

Figure 7.2. In addition to transferring kinetic energy to a planetary surface, some of the energy deposited in a meteor impact can go toward melting and other mineral phase changes in the target material. Left: Two tektites representing ejecta that melted during a large impact on the Earth's surface, and cooled while still in flight. Image credit: This image of two tektites has been obtained by the author from the Wikimedia website where it was made available by Brocken Inaglory under a CC BY-SA 3.0 licence. It is included within this article on that basis. It is attributed to Brocken Inaglory. Right: Phase diagram for SiO_2. All labels besides "Liquid" refer to different solid-phase polymorphs. While SiO_2 is most commonly found as quartz on Earth's surface, the high pressure forms stishovite and coesite are commonly found at impact sites. Data from Mao et al. (2001).

impact sites. (The lower right portion of the plot on the other hand is hot but at lower pressure, and these polymorphs of quartz typically occur in volcanic environments.) Since stishovite and coesite are found at impact sites, some of an impactor's energy is therefore being used to convert quartz into higher-pressure phases. Given these pieces of evidence we can say that by considering only gravity in Equation (7.4) we have derived just an approximation of the energy needed to form a crater. Nevertheless, since we have shown that the gravity term scales roughly as R^4, it is likely to dominate over terms more directly proportional to the volume of rock ejected (like fracturing, melting, and mineral phase changes), which will grow more slowly as R^3.

Equipped with Equation (7.4) we can do some calculations of the amount of energy required to create some well-known craters on Earth. One such crater is the aptly named Meteor Crater (also known as Barringer Crater), located in northern Arizona. It has a radius of about 0.6 km. For simplicity, we'll assume that the crust had a density, ρ_{crust} of 3000 kg m^{-3}. With these values, then, the impactor that created it must have had an energy of

$$KE \approx \frac{2\pi}{3}(3000 \text{ kg m}^{-3})(9.8 \text{ m s}^{-2})(6 \times 10^2 \text{ m})^4 \tag{7.6}$$

$$\approx 8 \times 10^{15} \text{ J}. \tag{7.7}$$

To get a feel for what this implies, here are a few facts for comparison. The caloric energy in an energy bar is about 10^3 J, so this is obviously a different scale of energy compared to what we are used to in everyday life. The largest nuclear bomb ever tested by humans released roughly 10^{17} J of energy, so the Meteor Crater impact comes much closer in scale to that explosion, which was reported to have broken windows up to 900 km away from the blast. We can also simplistically estimate whether this impact might have had a large effect on Earth's climate, by comparing to the energy received by the Earth from the Sun over a year, which is about 5×10^{24} J. The impactor had much less energy than this, so we can hypothesize that it is unlikely to have affected Earth's climate. It can also be interesting to compare to the energy of Earth's orbit, which is given in Equation (3.57). If an impact is to significantly alter a planet's orbital trajectory, it must have an energy comparable to that of the planet's orbit. For the Earth,

$$E_{orb} = -\frac{GM_\odot M_\oplus}{2a} \tag{7.8}$$

$$\approx -3 \times 10^{33} \text{ J}. \tag{7.9}$$

Therefore, although the impact was quite energetic from a human perspective, it would not have had a noticeable effect on the Earth's orbit.

Another well-known crater is the Chicxulub crater, located beneath the Yucatán peninsula, and believed to be the remnant of the impact that caused the extinction of the dinosaurs. The radius of this crater is about 90 km. To estimate the energy of this impactor we'll practice using *proportional reasoning* (first discussed in Section 2.2), using the proportional relationship in Equation (7.5) in place of the exact Equation (7.4). Remember that proportional reasoning allows us to estimate values without knowing the exact form of an equation, but only knowing how the desired value scales with a key parameter. Given Equation (7.5), we can say that $KE = CR^4$, where C is some constant that we don't need to know. Then, using the subscripts C and M to represent the Chicxulub and Meteor Crater impacts, the ratio of their impactor kinetic energies is given by:

$$\frac{KE_C}{KE_M} = \left(\frac{R_C}{R_M}\right)^4. \tag{7.10}$$

Notice how the constant cancels out when we take the ratio of kinetic energies, making its value unnecessary for the calculation. Therefore,

$$KE_C = \left(\frac{R_C}{R_M}\right)^4 KE_M \tag{7.11}$$

$$= \left(\frac{90 \text{ km}}{0.6 \text{ km}}\right)^4 \times 8 \times 10^{15} \text{ J} \tag{7.12}$$

$$= 5 \times 10^8 \times 8 \times 10^{15} \text{ J} \tag{7.13}$$

$$= 4 \times 10^{24} \text{ J}. \tag{7.14}$$

Recall that this is similar to the amount of energy reaching the Earth in a year. Therefore, one could imagine that the delivery of this much energy to the Earth's surface has the potential to affect its climate, which can help explain why this impact could have significantly altered the ecosystems on Earth (and killed off the dinosaurs). Even this large of an impact, however, would not have affected Earth's orbit appreciably.

7.1.2 Cratering as a Probe of Geologic History

The presence of craters on a surface can also be used as a probe of the body's geologic history. Imagine a planetary surface as a sandbox that has just been leveled out. Then imagine that the sandbox is subsequently bombarded with marbles at some regular time interval, perhaps one marble per minute. If you observe the sandbox and notice that there are 5 craters present, you know that it has been 5 min since the sandbox was leveled. Or, if you see 100 craters, it has been 100 min since the sandbox was leveled. In this same way, we can use cratering to figure out the age of a planetary surface. If a surface is heavily cratered, it has been a long time since that surface formed. A surface with few or no craters has been recently formed or resurfaced by geologic processes such as volcanism (as is the case for Io) or erosion (as often happens on the Earth).

If we consider the surface features of planetary bodies in the solar system, we'll notice that it is generally the larger bodies (especially Earth and Venus) with the fewest craters, indicating ongoing geologic activity. Smaller bodies like the Moon and Mercury are heavily cratered, indicating a lack of recent geologic activity, and Mars is intermediate, with evidence for a range of different surface ages indicating declining geologic activity over time. Figure 7.7 shows that the difference in crater abundance between these bodies is evident even at the global scale. There are also a few notable exceptions to this trend, such as the Galilean satellites Io and Europa, which are similar in size to Earth's Moon, but very geologically active. As we discussed in Chapter 6, geologic activity requires the presence of internal heat. For large, geologically active, planets, this heat is provided during the formation process as well as by radioactive elements. The reasons these smaller bodies are able to remain geologically active, is typically due to an alternate and ongoing energy source (usually tidal interactions) and/or a difference in composition (icy materials in particular can flow and melt at much colder temperatures than silicate rocks).

Saturn's largest moon Titan is another interesting exception to this trend, retaining relatively few craters on its surface. Titan retains a thick, nitrogen-rich atmosphere (see Figure 7.3), and hosts a methane-based weather cycle, analogous to Earth's hydrological cycle (Hayes et al. 2018). The left-hand side of Figure 7.3(left) shows methane river channels on Titan's surface as revealed by the Huygens lander, which was part of the Cassini mission to Saturn.

Pluto provides another interesting exception to the basic trend of small bodies being geologically inactive. Although chosen as the target for NASA's New Horizon space mission, many scientists predicted that, being both small and cold, Pluto was likely to be heavily cratered, and geologically unremarkable. Instead, Pluto turned out to host some surface regions that were nearly completely devoid of craters. Because the surface temperature on Pluto is close to the freezing point of N_2, nitrogen can transition easily between the solid and vapor phases, and can flow over time like water ice glaciers on the Earth (Moore et al. 2016). In addition, Pluto's large orbital eccentricity causes the solar radiation received by the planet to change

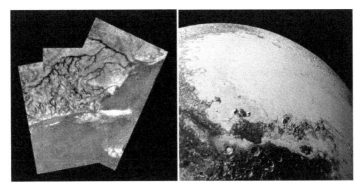

Figure 7.3. Two examples of young geologic surfaces in the outer solar system. Left: Image of river channels on Titan's surface as seen by the Huygens lander during its descent to the Moon's surface, showing evidence for flowing liquids. Image source: NASA/JPL/ESA/University of Arizona. Right: False color image of Sputnik Planitia on Pluto, composed of solid nitrogen ice, which is able to flow like a glacier at Pluto surface temperatures. Image source: NASA/Johns Hopkins Applied Physics Lab/Southwest Research Institute.

significantly over the course of its 248 Earth-year long orbit, resulting in seasonal exchange of nitrogen between its surface and thin atmosphere (Bertrand et al. 2018). The most prominent solid nitrogen deposit, Sputnik Planitia, is shown in Figure 7.3 (right).

7.1.3 Craters as a Probe of Solar System History

Craters can also be used to study the impact history of the solar system. So far, we've simply discussed how older surfaces have more craters, while younger surfaces have fewer. But suppose we can determine the age of a surface via some other means. In that case, it would be possible to infer the rate of impacts. It may help to think back to the sandbox analogy—if we observe five craters, and also know that it has been 5 min since the sandbox was leveled, we can infer that one crater is created per minute. If we had many surfaces with different ages, it would also be possible to test whether the impact rate is constant in time, or if it has instead changed throughout the solar system's history.

How else might one obtain the age of a planetary surface? If one can actually collect rocks from the surface, it's possible to use *radiometric dating* to determine when the rock was formed. Radiometric dating uses the amounts of parent and daughter elements, plus the parent's half-life, to estimate the age of a rock; we touched on this briefly in Section 6.3.2 but will revisit it in more detail in Section 9.6.3. While it is obviously possible to collect rocks on Earth, Earth's cratering record has been mostly erased by geological activity and plate recycling (which we'll discuss further in Section 7.2). However, we also have samples of rocks from the Moon's surface, thanks to the Apollo missions. Before discussing what we can learn from these data, it's important to discuss two caveats about crater dating of surfaces. First, if a large enough number of impacts have occurred, and a large enough number of craters has been created, it may become the case that the creation of any new crater on average erases the evidence of one old crater. This phenomenon, known as *crater saturation*, is demonstrated in Figure 7.4. Notice how the first frames of this simulation show the number of craters growing with time, while the last few frames are difficult to tell apart, even though impacts have not ceased. Simple geometric reasoning tells us that there are a maximum number of craters of a given size that can fit into any given area—therefore, some caution must be utilized when considering the age of a heavily cratered surface.

A second complication is that, unlike in our simple sandbox analogy, the meteorites hitting the surface are going to have a range of masses and kinetic energies, and create a range of crater sizes. Based on intuition alone, we might expect that there are more small meteorites than large, and this intuition is correct. In fact, the production of space debris occurs in a process known as a collisional cascade in which all collisions result in smaller products, such that the smallest objects become most numerous. Therefore, rather than count total numbers of craters, it's important to note both the number and size of the craters being counted, and to compare two surfaces' crater counts as a function of crater size.

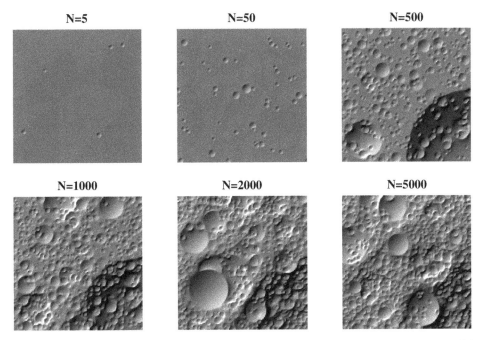

Figure 7.4. Snapshots of a simulated surface accumulating an increasing number of craters (N), based off the model of Rosenburg et al. (2015). Although the first few frames are easy to distinguish in terms of crater abundance, the last few frames are more difficult to tell apart, since the surface has started to become saturated.

Figure 7.5 shows actual crater size distribution data collected for surfaces on the Moon and Mars. Notice how the curves all show more small craters created than large craters, and how the total number of craters for each size bin increases with surface age. The absolute ages for the Moon come from dating of rocks on the Moon's surface, while those for Mars come from the lunar ages, adjusted for the expected difference in impact rates for Mars as compared to the Moon.

Figure 7.6 shows the cumulative crater density of different surfaces on the Moon versus their radiometric age, giving us a reconstructed view of the history of cratering. A straight line on this logarithmic plot would indicate that craters are produced at a constant rate throughout the Moon's geologic history. While this is consistent with the Moon's recent history, the data show a surge in crater formation at the earliest times in the solar system's history. As we'll discuss in Chapter 9, collisions were frequent in the early solar system, but decreased with time as planets grew.

Cratering studies have also revealed an interesting phenomenon for surfaces created between about 3.8 and 3.9 billion years before the present day. These surfaces show an enhanced amount of impact features, suggesting a short-lived uptick in cratering that has come to be called the *Late Heavy Bombardment*. Explanations for the cause of the apparent Late Heavy Bombardment have been actively discussed in the context of solar system formation histories. We will revisit this when we discuss the so-called "Nice model" in Chapter 9.

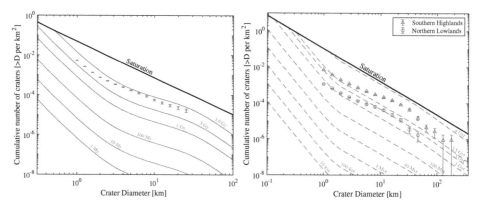

Figure 7.5. Left: An example of cratering abundance as a function of size for the Moon, from the Robbins crater database (Robbins 2019). Red points show actual data, while gray lines show expected curves for different surface ages, called "isochrons". The solid black line shows the expected number of craters of each size if the surface is saturated with craters. Right: Crater data for Mars's northern and southern hemispheres, from Robbins & Hynek (2012). The Northern hemisphere has significantly fewer craters than the south, implying a younger surface age.

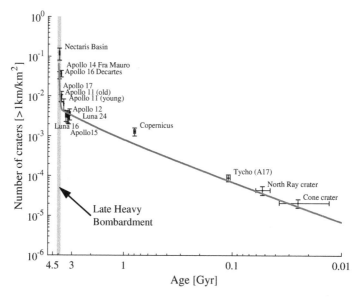

Figure 7.6. Crater density plot for the Moon, demonstrating a non-linear rate of crater formation. The x-axis is given in units of "Gyr", which means Giga-years, or billions of years before the present day. Data from Stöffler & Ryder (2001), model from Neukum et al. (2001).

7.2 Geologic Activity on the Terrestrial Planets

Armed with an understanding of how to use cratering to date surfaces, we can very briefly discuss some basic geological characteristics of the terrestrial planets— Mercury, Venus, Earth, and Mars.

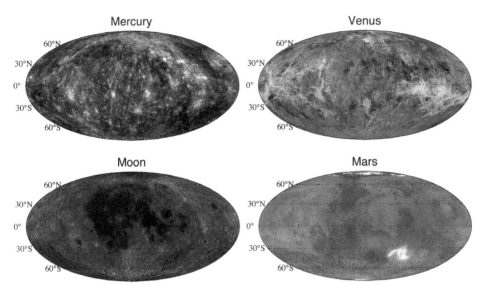

Figure 7.7. Global mosaics of three terrestrial planets and the Moon. Mercury: Enhanced color mosaic from the MESSENGER MDIS camera (Johns Hopkins Applied Physics Laboratory). Venus: False color Magellan RADAR mosaic (USGS). Moon: Grayscale mosaic from the Lunar Reconnaissance Orbiter Wide Angle Camera (Arizona State University). Mars: Approximately true color Viking image mosaic (USGS).

Mercury is heavily cratered across its entire surface, as shown in Figure 7.7, suggesting an overall lack of recent geologic activity. Among the more prominent geologic features aside from cratering are compressional tectonic features known as "lobate scarps", distributed across much of the planet. These features provide evidence that the planet has contracted radially as it has cooled over the solar system's history, causing the crust to undergo faulting to accommodate the reduced surface area. Like the other terrestrial bodies in the inner solar system other than the Earth, the crust of Mercury is not broken up into separate plates, but is considered to be a single plate. In many respects Mercury and the Moon, as smaller airless bodies, exhibit similar crater-dominated surface geology (Figure 7.7).

Venus, on the other hand, has only a small number of craters on its surface. This difference arises for two reasons. For one, Venus has a thick atmosphere (while Mercury, in contrast, is essentially airless). In fact, Venus's atmosphere is so thick that it prevents visible-light investigations of its surface from space, and so the primary information about Venus's surface has come from radio-wavelength *RADAR* mapping (see Section 7.3.2), especially from the Magellan mission (Figure 7.7). The thick atmosphere of Venus prevents many meteorites from reaching the ground, either slowing them down or causing them to burn up entirely. However, this effect is primarily applicable for smaller craters, and doesn't explain the lack of larger craters as well. Therefore, secondarily, Venus must have undergone resurfacing to destroy most of the larger craters that would have formed in its past. The number of existing craters on Venus suggests that it was nearly completely resurfaced somewhere in the range of 300–750 Million years (My) ago. Although a long time ago from a human perspective, this is a recent event in terms of geologic timescales. The exact mechanism of this resurfacing is not known, but it does suggest that volcanic or tectonic activity was able to bury or destroy craters. The large number of volcanoes and tectonic features on Venus indicate a geologically active planet, likely continuing to the present day to some extent. However Venus rejuvenates its crust, it is clear that this must happen in different ways than on the Earth.

Earth is also a geologically active planet, with an average surface age of order 100 Myr. In contrast to Venus, the visible geologic history of Earth is dominated by a process known as *plate tectonics*. Heat from the Earth's interior drives convection in the Earth's mantle. (Note that although the mantle is solid throughout, it is able to flow over long timescales, like a glacier.) These convective motions in turn help to drive movement of Earth's crust, which is broken into a number of separately moving plates. These plates are composed of two distinct types of crust—thicker, more buoyant continental crust, and thinner, denser oceanic crust. Many of the familiar geological features of the Earth occur at plate boundaries. Collisions between continental plates can produce long mountain belts like the Himalayas or Appalachians. Subduction of an oceanic plate (when one plate dips below another and sinks back into the underlying mantle) often produces chains of volcanoes along the plate margin, like the Andes, and subduction zones are the source of some of the Earth's highest-energy earthquakes. Where two plates diverge, such as at the mid-Atlantic ridge, a rift or trench forms, and new oceanic crust can be created by upwelling lava. Over time, collision and subduction of plates leads to recycling of the crust back into the mantle, balanced by the creation of new crust via volcanic activity.

Exactly why Earth developed plate tectonics while Venus did not is still under debate. With similar masses, sizes and compositions, Venus and Earth should have similarly active interiors. There is evidence that liquid water is a necessary ingredient for the development of plate tectonics, as it facilitates melting and plate motion. Venus is a water-poor planet compared to Earth, and its surface temperature is so high that it cannot currently support liquid water on its surface. Both of these facts are likely a consequence of Venus's closer distance to the Sun. However, as we will discuss in Chapter 8, Venus's surface temperature (about 735 K) is much higher than its equilibrium temperature; therefore, it may be the development and retention of its thick greenhouse atmosphere that caused a large divergence in its geologic properties from those of Earth.

Mars's geology is dominated by a distinct *dichotomy* between its northern and southern hemispheres, which can be seen in the topographic elevation map shown in Figure 7.8. The northern hemisphere has surface features whose heights average 1–3 km lower than the southern hemisphere. The southern highlands are relatively crater-rich, suggesting that the surface is old (recall Figure 7.5). The northern lowlands are relatively crater-poor, and the boundary between the two regions has a large number of features associated with long-standing surface water. Therefore, the northern lowlands may have been more recently shaped by geologic activity and/or liquid erosion. Two leading hypotheses for the creation of the Martian dichotomy

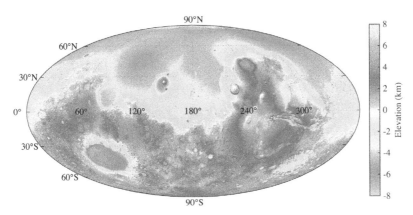

Figure 7.8. Topographic map of Mars obtained with the Mars Orbiter Laser Altimeter (MOLA), showing the clear dichotomy between the smoother northern lowlands and the more heavily cratered southern highlands.

are a giant impact, and degree-1 mantle convection. In a giant impact scenario, a large impactor hits the northern hemisphere, creating a nearly hemisphere-wide crater (Andrews-Hanna et al. 2008). In the degree-1 mantle convection scenario, the mantle undergoes a mode of convection with upwelling in one hemisphere and downwelling in the other hemisphere (Roberts & Zhong 2006). This convection pattern sets up different thermal conditions in the two hemispheres that could lead to differences in crustal thickness, and thus surface elevation.

Mars also exhibits a wide range of features on its surface indicative of a geologically rich past. Evidence of past rivers and lakes are widespread across portions of its surface, and have been an ongoing target of exploration to understand the past habitability of the planet. Mars hosts the largest volcanoes in the solar system, particularly in the elevated Tharsis region of the planet (look for the white peaks near 240° longitude in Figure 7.8), along with extensive tectonic features indicative of past activity. Meanwhile, modern glaciers, ice caps, and wind-blown sand dunes demonstrate the ongoing geologic activity of the Martian surface.

7.3 Surface Composition

Studies of the chemical composition and mineralogical make up of planetary surfaces can reveal an abundance of detailed information about geological processes, interactions between the surface, atmosphere, and hydrosphere (water at the surface or below ground), and even the effects of life—information we can by no means do justice to in this text. Instead, we'll think very broadly about what elements and minerals make up a planet's crust, and how we can determine these properties.

7.3.1 Factors That Determine Surface Composition

Ultimately, the composition of the planets are determined in large part by the original distribution of elemental abundances in the early solar system (see Figure 6.5 from Chapter 6), as well as the local temperatures where each planet forms As we will discuss further in Chapter 9, the temperature in the early solar system falls off with distance from the young Sun, allowing different types of elements to condense as solids at different locations. For this reason, planetary surfaces in the inner solar system are depleted in so-called volatile elements like H, C, O, and N (which would have existed in the vapor phase), while surfaces in the outer solar system are dominated by such elements (as they were cold enough to exist in the solid phase). Outer solar system bodies therefore have surfaces dominated by ices made of those elements, like water, ammonia, carbon mono- and dioxide, and even (at very low temperatures) solid nitrogen. In the inner solar system, these ices were not common in solid form, and the building blocks of rocky planetary crusts are instead largely silicate minerals and metal oxides. The most important rock-forming elements in the Earth's crust include O, Si, Al, Fe, Ca, Na, K, and Mg (Figure 6.5). These elements make up a number of major mineral groups found in the crusts of terrestrial planets, including pyroxenes, amphiboles, feldspars, micas, and clays.

7.3.2 Remote Sensing of Planetary Surfaces

On the Earth, we have a variety of analytical techniques at our disposal to determine the mineral or chemical composition of a rock. However, for most bodies in the solar system, we must utilize *remote sensing* techniques to study surface composition. For compositional analyses, remote sensing primarily utilizes the interaction of *photons* (particles of light) with the surface of a planet, detected either by Earth-based telescopes or spacecraft-mounted instruments. The most important interactions of photons with a surface include reflection, absorption, and emission. Spectroscopy, the measurement of numbers of photons (synonymous with brightness) as a function of wavelength (also discussed in Section 5.2.3), can be used in combination with each of these processes at a variety of wavelengths across much of the electromagnetic spectrum. We illustrate the wavelength ranges of different types of spectroscopy in Figure 7.9. Photons of different energy can excite, or be emitted by, a range of atomic and molecular transitions, as shown in Figure 7.10 (left). Because all of these transitions are *quantized* (restricted to discrete energy levels), and unique to a given atom or molecule, they can provide diagnostic signatures of a planet's composition in its reflected or emitted spectrum.

The energy of a photon is given by $E = h\nu$, where ν is the frequency of the light, and h is Planck's constant (for this reason, we use $h\nu$ to represent the energy of an emitted or absorbed photon in Figure 7.9). In addition, the wavelength and frequency of a photon are inversely related by the relationship $\lambda = \frac{c}{\nu}$, where λ is the wavelength and c is the speed of light. Therefore, different wavelengths probe different types of energy regimes in atoms and molecules. In addition, high frequencies correspond to short wavelengths, while low frequencies correspond to long wavelengths.

At the highest frequency end of the electromagnetic spectrum, gamma-rays are able to probe the highest energy atomic transitions. Transitions between energy levels in atomic nuclei dominate this region of the spectrum, which is why

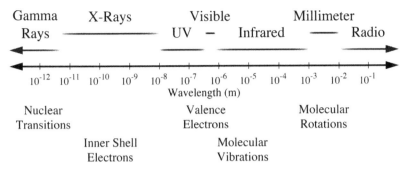

Figure 7.9. Map of the electromagnetic spectrum, from gamma-rays (highest energy) to radio waves (lowest energy). Photons of different energy are capable of exciting, or being emitted by, a range of atomic and molecular transitions that can be used by spectroscopic techniques to determine the chemical and mineralogical composition of a planet's surface.

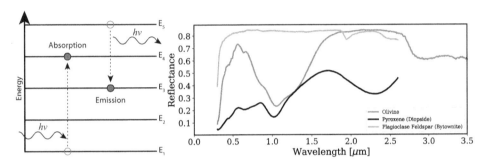

Figure 7.10. Left: Spectroscopy provides diagnostic information about a planet's composition due to the fact that energy levels (E_i) in an atom or molecule are quantized. The atom or molecule absorbs or emits characteristic wavelengths of light (or, in other words, photons with defined energies) when it transitions between energy levels. Right: Reflectance spectra of three common mineral groups found on Earth—olivine, pyroxene and plagioclase feldspar, illustrating the diagnostic information present in the visible and near-infrared portions of the spectrum. Reflectance is measured as a ratio relative to the incoming source spectrum, from 0 (completely absorbing) to 1 (completely reflecting). Data from RELAB Spectral Database taken by E. A. Cloutis and C. M. Pieters.

radioactive nuclei are one common source of gamma-rays. Another source of nuclear excitation in this region of the spectrum are high-energy cosmic rays that continually bombard a planet's surface (for bodies with thin or no atmospheres). As a result, gamma-ray spectroscopy can be used to measure and map the abundance of different elements on a planet's surface, and has been used to great effect at the Moon, Mercury, Mars, and other bodies with thin or no atmospheres.

At lower frequencies, electron transitions between different energy levels within an atom's electron shells become more dominant. The X-ray and ultraviolet (UV) regions of the spectrum are not commonly used for studying planetary surfaces, which tend to absorb most of the incoming light in this range, rather than reflecting it. However, some electronic transitions, particularly among the outer shell (valence) electrons of atoms, occur in the visible portion of the spectrum. These transitions are responsible for many of the colors we can appreciate in rocks with our own eyes. One notable example is the red hue of Mars, which is due to abundant iron oxides on its surface. Another example is the difference in brightness on the Moon between the lighter highlands and the darker patches known as "maria" (see Figure 7.7). The basalt rocks that make up the maria are comparatively rich in elements like iron and titanium that absorb light strongly in the visible spectrum, causing a darker color. Even small amounts of elements like iron, titanium, chromium, or copper can cause dramatic changes to the color of a mineral. In addition, an abundant source of visible photons is available in the form of sunlight, which can be absorbed or reflected by a planet's surface. This is the basis of reflectance spectroscopy, where the ratio between the incoming solar energy spectrum and the spectrum reflected off of a planet provide information about the composition of the surface. Figure 7.10 shows example reflectance spectra of several minerals in the visible and near-infrared portions of the electromagnetic spectrum.

Moving farther into the infrared, transitions between energy levels defined by vibration of atoms within larger molecules or crystal lattices dominate the spectral signature of planets. These transitions require lower energy to be excited than the electronic and nuclear transitions discussed above, and generally fall in the 1–50 μm wavelength range. Recalling the discussion of blackbody radiation in Chapter 2, and in particular Figure 2.2, you will notice that this range coincides quite well with the peak in emission for a planetary surface temperature on the order of a few hundred degrees Kelvin. Therefore, rather than observing reflected solar radiation, emission spectroscopy from the thermal radiation of the planet itself can be used. Spectral features in this region can provide diagnostic information about the mineralogy of a planet's surface, using the energy level transitions of key mineral components like the Si–O, or O–H bonds, for example.

At even lower energy far-infrared and millimeter wavelengths, molecular rotational energy transitions are present, but are generally not important for solid planetary surfaces where molecules are not free to rotate. Finally, at the very longest wavelengths in the radio spectrum, energies are generally too low to excite diagnostic quantum transitions. However, this fact also enables radio waves to propagate easily even through thick atmospheres like on Venus or Titan, allowing us to obtain information about the surfaces of those planets. For this reason, RADAR is a commonly used technique for remote sensing of planetary surfaces. In the RADAR technique (which stands for RAdio Detection And Ranging), radio waves are transmitted from a spacecraft and reflected off interfaces (like a planet's surface) where there is a change in relative permittivity (also known as the dielectric constant). Relative permittivity, which you may have encountered in an Electricity and Magnetism physics course, is a measure of how a material responds to an electric field. The scattering of radio waves off of these interfaces can provide information about the relative permittivity of the material as well as the macroscopic roughness of the interface, providing indirect information about surface composition (and topography). The global mosaic of Venus in Figure 7.7 is one example of the use of radio waves in remote sensing.

7.4 Important Terms

- tectonic activity;
- impact ejecta;
- radiometric dating;
- crater saturation;
- Late Heavy Bombardment;
- RADAR mapping;
- plate tectonics;
- Martian dichotomy;
- remote sensing;
- photon;
- quantized energy levels.

7.5 Chapter 7 Homework Questions

1. **Crater counting:**[1]

   In this question, we'll create a very simple crater simulation, to investigate the basics (and potential pitfalls) of crater age dating. You will create simulated maps of 48 craters on a planetary surface. To create these maps, you will need to use a computer programming language to generate random numbers, and make plots. Divide your maps into an eight element wide by six element wide grid, and use a random number generator to select the crater locations. Plot the craters on the grid by placing a circle at each (x, y) position generated by your random number generator.

   (a) Produce the following six models, and display their plots:
   - Model 1: Evenly spaced
     Put one crater at a random location in each cell (one crater per cell).
   - Model 2: Random
     Place the 48 craters at random locations on the grid.
   - Model 3: 40% Enhanced
     Place 28 craters at random on the right half of the grid and 20 craters at random on the left half of the grid.
   - Model 4: 100% Enhanced
     Place 32 craters at random on the right half of the grid and 16 craters at random on the left half of the grid.
   - Model 5: 200% Enhanced
     Place 36 craters at random on the right half of the grid and 12 craters at random on the left half of the grid.

---

[1] Adapted from de Pater & Lissauer (2010) and lecture notes of F. Chromey.

- Model 6: 400% Enhanced
    Place 40 craters at random on the right half of the grid and 8 craters at random on the left half of the grid.
    Based on your maps, comment on the following questions:
(b) How does the evenly spaced model compare to the random models? Were you surprised by how a "random" surface looks? If so, how? Comment on the tendency to perceive structure in the random maps.
(c) How easy is it to identify the two cratering rates in these maps (keeping in mind that on a real planet, you would have no way to know ahead of time where the division sits)? How might you caution an investigator using cratering maps?
(d) Given these models, what strategies might you use to get accurate measurements of (relative) surface ages?
(e) Describe one or more ways you might make a computer-based crater simulation more realistic.

References

Andrews-Hanna, J. C., Zuber, M. T., & Banerdt, W. B. 2008, Natur, 453, 1212

Bertrand, T., Forget, F., Umurhan, O. M., et al. 2018, Icar, 309, 277

de Pater, I., & Lissauer, J. J. 2010, Planetary Sciences (Cambridge: Cambridge Univ. Press)

Hayes, A. G., Lorenz, R. D., & Lunine, J. I. 2018, NatGe, 11, 306

Mao, H., Sundman, B., Wang, Z., & Saxena, S. K. 2001, JAlIC, 327, 253

Moore, J. M., McKinnon, W. B., Spencer, J. R., et al. 2016, Sci, 351, 1284

Neukum, G., Ivanov, B. A., & Hartmann, W. K. 2001, SSRv, 96, 55

Roberts, J. H., & Zhong, S. 2006, JGRE, 111, E06013

Robbins, S. J., & Hynek, B. M. 2012, JGRE, 117, E05004

Robbins, S. J. 2019, JGRE, 124, 871

Rosenburg, M. A., Aharonson, O., & Sari, R. 2015, JGRE, 120, 177

Stöffler, D., & Ryder, G. 2001, SSRv, 96, 9

Introductory Notes on Planetary Science
The solar system, exoplanets and planet formation
Colette Salyk and Kevin Lewis

Chapter 8

Planetary Atmospheres

8.1 Escape Speed and Its Effect on Planetary Atmospheres

8.1.1 Broad Properties of Solar System Atmospheres

If we think very broadly about the atmospheres of planets in the solar system, we'll notice that their bulk properties are correlated with planet mass. The most massive bodies in the solar system (the gas and ice giants) have very massive atmospheres composed of a large fraction of light elements like hydrogen and helium. Medium-sized bodies, like the Earth, Venus, and Mars, have much smaller atmospheres, lacking significant quantities of hydrogen and helium. The smallest bodies in the solar system, including Mercury, or the asteroids, have either non-existent or very tenuous atmospheres known as *exospheres*. Exospheres are composed of elements either released from the planet's surface or delivered by the solar wind or meteorites. By the end of this section, we'll understand why planet mass is linked to atmospheric size and composition.

8.1.2 The Physics of Escape Speed

The *escape velocity* or *escape speed* of a planet is the minimum speed an object needs to escape the gravitational pull of the planet.[1] (Although this quantity is often called escape *velocity*, it has no direction associated with it and so is more properly called escape *speed*. However, you will find both terms used interchangeably. In this text, we refer to it as speed, and give it the symbol v_{esc}.) An analogous concept often discussed in introductory physics courses is the minimum speed a ball would need at the bottom of a hill to make it to the top of the hill. As shown in Figure 8.1, in this common physics problem, to find the *minimum* speed the ball requires to crest the hill, we assume that when the ball reaches the top of the hill, at height h, it has a final speed, v_f of 0. Equating the total energy (kinetic plus potential) of the ball at the bottom and top of the hill allows us to find the speed at the bottom, v_i:

---

[1] If it has nothing pushing it along.

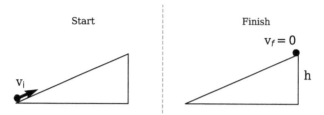

Figure 8.1. Setup for a common problem discussed in introductory Physics classes—what is the minimum speed (v_i) the ball would need to have at the bottom of the hill if it's to reach the top of the hill without stopping? Such a problem is analogous to finding the escape speed of an object from the surface of a planet.

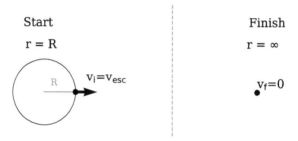

Figure 8.2. Setup for finding the escape speed from a planet. The object begins at the surface of the planet ($r = R$) with some speed v_i, and at a far distance from the planet ($r = \infty$) it has a speed of 0.

$$E_i = E_f \tag{8.1}$$

$$U_i + K_i = U_f + K_f \tag{8.2}$$

$$0 + \frac{1}{2}mv_i^2 = mgh + 0 \tag{8.3}$$

$$v_i = \sqrt{2gh}. \tag{8.4}$$

For a planet, we can ask, what is the minimum speed an object would need at the surface of the planet of mass M and radius R if it is to eventually reach some very far distance from the planet? In this case, as shown in Figure 8.2, the object begins at the surface of the planet with position $r = R$, and at a very large distance from the planet, $r = \infty$, it has a speed, v_f equal to 0. Equating the total mechanical energy of the object at the planet's surface and far from the planet allows us to find v_i, which is the escape speed, v_{esc}.

$$E_i = E_f \tag{8.5}$$

$$U_i + K_i = U_f + K_f \tag{8.6}$$

$$-\frac{GMm}{R} + \frac{1}{2}mv_i^2 = -\frac{GMm}{\infty} + 0 \tag{8.7}$$

$$v_i = \sqrt{\frac{2GM}{R}} \qquad (8.8)$$

$$v_{esc} = \sqrt{\frac{2GM}{R}}. \qquad (8.9)$$

To get a feel for escape speed, let's compute its value for Earth.

$$v_{esc,\oplus} = \sqrt{\frac{2GM_\oplus}{R_\oplus}} \qquad (8.10)$$

$$= \sqrt{\frac{2(6.67 \times 10^{-11}\ \text{m}^3\ \text{kg}^{-1}\ \text{s}^{-2})(5.972 \times 10^{24}\ \text{kg})}{6.371 \times 10^6\ \text{m}}} \qquad (8.11)$$

$$= 11.2\ \text{km s}^{-1}. \qquad (8.12)$$

This number is (not surprisingly!) much higher than everyday speeds, such as a 60 mph car ($v = 0.027$ km s^{-1}) or even a 500 mph airplane ($v = 0.22$ km s^{-1}). However, interplanetary spacecraft have launch speeds higher than $v_{esc,\oplus}$, so that they can escape the Earth's pull and travel to other parts of the solar system.

Since escape speed depends on both M and R it can be instructive to see how it scales with planetary mass if we assume a constant density for all planets. Since

$$v_{esc} \propto \left(\frac{M}{R}\right)^{1/2} \qquad (8.13)$$

and

$$R \propto M^{1/3} \qquad (8.14)$$

if all planets have the same density, then

$$v_{esc} \propto \left(\frac{M}{M^{1/3}}\right)^{1/2} \qquad (8.15)$$

$$\propto (M^{2/3})^{1/2} \qquad (8.16)$$

$$\propto M^{1/3}. \qquad (8.17)$$

Escape speed is larger for more massive planets. Therefore, particles in a planet's atmosphere would need to be moving more quickly if they are to escape a massive planet than if they are to escape a small, low-mass planet. But what speeds *do* particles in a planet's atmosphere have? We will discuss this in the next section.

8.1.3 Thermal Speeds, Atmospheric Escape, and Atmospheric Composition

For ideal gases, the average kinetic energy of gas molecules is given by $K_{avg} = \frac{3}{2}kT$. Using the definition of kinetic energy, we can use this fact to derive the average speed of a molecule of mass m in an atmosphere at temperature T.

$$K_{avg} = \frac{1}{2}mv_{avg}^2 \tag{8.18}$$

$$\frac{3}{2}kT = \frac{1}{2}mv_{avg}^2 \tag{8.19}$$

$$v_{avg} = \sqrt{\frac{3kT}{m}}. \tag{8.20}$$

As an example, Earth's atmosphere is primarily composed of N_2, which has a mass of 28 atomic mass units, or 4.6×10^{-26} kg. Thus, the average speed of N_2 molecules in Earth's atmosphere is about:

$$v_{avg} \approx \sqrt{\frac{3 \times 1.38 \times 10^{-23} \text{ m}^2 \text{ kg s}^{-2} \text{ K}^{-1} \times 300 \text{ K}}{4.6 \times 10^{-26} \text{ kg}}} \tag{8.21}$$

$$\approx 520 \text{ m s}^{-1}. \tag{8.22}$$

This is much slower than the escape speed, which suggests that Earth's N_2-rich atmosphere is unlikely to escape the Earth's gravitational pull. However, we have so far only found the *average* speed of molecules. In fact, the speed of a collection of identical molecules at a single temperature is described with a probability distribution known as the *Maxwell–Boltzmann distribution*:

$$f(v) = \left(\frac{m}{2\pi kT}\right)^{3/2} 4\pi v^2 \exp^{-\frac{mv^2}{2kT}} \tag{8.23}$$

where m is the mass of the gas molecule or atom, T is the gas temperature, v is the speed, and k is the Boltzmann constant.

The distribution is shown graphically in Figure 8.3. As with all probability distributions (also known as probability density functions), the probability of finding an atom with a speed between two values (say, 1 km s^{-1} and 2 km s^{-1}) is given by the integral of the curve between those two values. While the average speed is near the peak of the distribution, and therefore represents a likely speed for a given molecule, there are many molecules with speeds lower than and higher than the average speed. However, if we consider N_2 molecules on Earth's surface at $T \approx 300$ K, it seems that even in the upper tail of the distribution, molecules are far from Earth's escape speed.

Lighter molecules, however, have higher escape speeds; for example, the average speed of H_2 molecules would be faster than that of N_2 molecules, by a factor of $\sqrt{28/2} \approx 3.7$. In addition, in Earth's uppermost atmosphere, temperatures are much

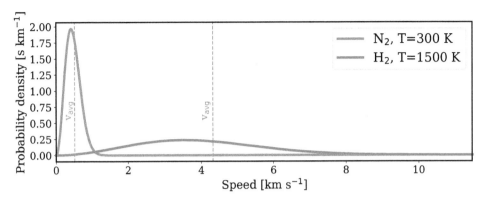

Figure 8.3. This figure shows the probability density function for the speed of N_2 gas at a temperature of 300 K, as well as H_2 gas at a temperature of 1500 K. The integral of the probability density function between two speeds v_1 and v_2 provides the probability of a molecule having a speed within that range. Less than $10^{-7}\%$ of the N_2 molecules at 300 K would have speeds greater than 2 km s^{-1}.

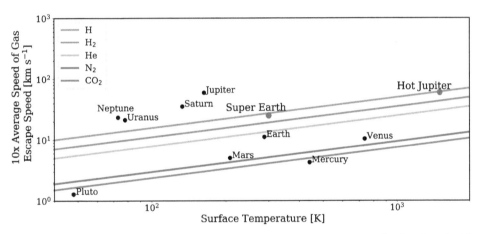

Figure 8.4. Plot summarizing the relationship between escape speed and gas speeds (at the planet's surface) for solar system planets, and for representative Super Earth and Hot Jupiter planets. Solid lines show 10× the average speed of the gas molecules as a function of temperature. Plot derived from The Nebraska Astronomy Applet Project (University of Nebraska, Lincoln 2020).

higher than they are at the surface. It turns out that, in fact, hydrogen does escape from the Earth at appreciable rates from the uppermost parts of Earth's atmosphere.

More massive planets, however, have an easier time retaining even light molecules, since their escape speeds are higher. For example, a Neptune-sized planet's atmosphere has an escape speed that is a factor of $\sqrt{\dfrac{M_N R_\oplus}{M_\oplus R_N}} = \sqrt{\dfrac{1 \times 10^{26}\text{ kg} \times 6300\text{ km}}{6 \times 10^{24}\text{ kg} \times 24600\text{ km}}} \approx 2$ times higher than Earth.

The relationship between escape speed and gas speed can be summarized as shown in Figure 8.4. The planets' temperatures and escape speeds are plotted along with solid lines showing 10× the average speed of different gas species as a function of temperature. The lines show 10× the average speed, rather than the average speed, as they're meant to account for the Maxwell–Boltzmann distribution, and the fact that some molecules will indeed have speeds higher than the average speed. The plot shows many features we've already discussed. The giant planets have higher escape speeds than the terrestrial planets. The speeds of light gases, like H_2, are higher than those of heavier molecules, like N_2. Whenever the gas speed line lies *above* the escape speed for a given planet, this implies that the gases are likely to escape. Thus if we look at Venus, Earth and Mars, we see that molecules like N_2 and CO_2 are retained, while lighter gases like H, H_2, or He are likely to escape. For the giant planets, however, the gas speeds are always below the escape speeds. Therefore, the giant planets can retain atmospheres full of light gases like H_2 and He. Mercury, however, lies below all of the gas speed lines, indicating that even heavier gases like N_2 and CO_2 will escape Mercury's gravitational pull.

This relatively simple physical picture does a remarkably good job of predicting planetary atmospheric compositions. The compositions of bodies in the solar system are shown in Figure 8.5. We can see that the terrestrial planets Mars, Venus and Earth retain only heavier gases, while the gas giants retain lighter gases like H and He. Mercury has an unusual atmospheric composition compared to the other terrestrial planets because the components of its tenuous atmosphere, known as an exosphere, are constantly escaping and being replenished by a variety of processes including high energy radiation hitting the planet's surface and freeing atoms into the atmosphere, release of atoms from the surface via sublimation or desorption, and implantation by the solar wind or meteorites. As these processes are location and time-dependent, the composition of Mercury's atmosphere cannot be easily shown in a simple pie chart. However, elements that have so far been detected in Mercury's atmosphere include H, He, Na, K, Ca, Mg, Al, Fe, Mn, and, possibly O (McClintock 2018).

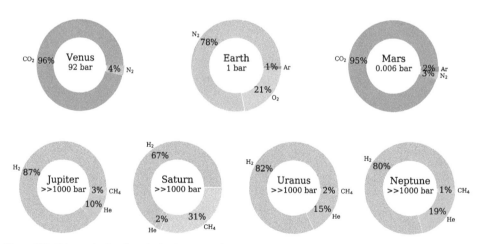

Figure 8.5. Schematic showing the basic composition of the solar system planet atmospheres. Inner labels show the atmospheric pressure, in bars. Compositions (% by volume) and pressures from the NASA Planetary Fact Sheet (NASA 2020). Mercury's atmospheric composition varies by position and time, and cannot be easily represented in this chart (e.g., McClintock 2018).

This simple picture also makes predictions for extrasolar planets—in particular, for hot Jupiters and super Earths. Figure 8.4 shows two points representing a hot Jupiter planet with an equilibrium temperature of 1500 K and Jupiter's mass and size, and a super Earth planet with mass of 10 $M_\oplus$ and radius $2R_\oplus$, and temperature of 300 K. The hot Jupiter has the same escape speed as Jupiter, but it's much hotter, so the lightest components of its atmosphere may escape at a non-negligible rate. The super Earth has the same temperature as Earth, but a higher escape speed; compared to Earth, it is more likely to have an atmosphere rich in light elements.

8.2 Atmospheric Structure

8.2.1 The Scale Height and Atmospheric Pressure

In this section, we'll discuss the vertical pressure structure of an atmosphere, and also understand why mountain climbers require supplemental oxygen. To begin this discussion, we imagine a small parcel of air of height Δz, cross-sectional area A, density ρ, and mass m, as shown in Figure 8.6. For a stable atmosphere, this parcel should feel zero net force, so that it neither falls nor sinks. Since the parcel feels a gravitational force downwards of magnitude mg, there must be some additional force equal in magnitude in the upwards direction for the parcel to remain balanced. This force is provided by the molecules at the bottom of the parcel pushing on the parcel with slightly more pressure than the molecules on the top of the parcel, and is called the *pressure gradient force*. Remember that pressure is force per unit area.

The force exerted upwards on the parcel has magnitude $F_p = p_{\text{bottom}}A - p_{\text{top}}A = -\Delta pA$, where p_{bottom} and p_{top} are the pressure exerted on the bottom and top of the parcel by the surrounding atmosphere, and $\Delta p = p_{\text{top}} - p_{\text{bottom}}$. To balance the gravitational force:

$$F_p = F_g \tag{8.24}$$

$$-\Delta pA = mg \tag{8.25}$$

$$-\Delta pA = \rho A \Delta z g \tag{8.26}$$

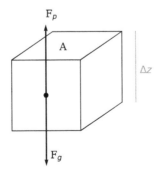

Figure 8.6. A small gas parcel in the atmosphere, from which we derive the equation for hydrostatic equilibrium.

$$\frac{\Delta p}{\Delta z} = -\rho g. \tag{8.27}$$

If we consider an infinitesimally small parcel, then:

$$\frac{dp}{dz} = -\rho g. \tag{8.28}$$

This is known as the *equation of hydrostatic equilibrium*. It represents the balance between the pressure gradient force and gravity.

If ρ and g were constant with height, this expression would tell us exactly how pressure changed with height. In fact, this is an excellent approximation for Earth's oceans, since water stays nearly the same density, even at the bottom of the ocean (we say that water is only slightly *compressible*), and g is nearly constant. For Earth's atmosphere, a constant g is a reasonable approximation, but a constant ρ is not. However, we can use the ideal gas law to relate p and ρ to derive an expression for $p(z)$. One form of the ideal gas law is:

$$p = \frac{\rho k T}{\mu} \tag{8.29}$$

where p is pressure, ρ is density, k is the Boltzmann constant, T is temperature, and μ is the mean molecular mass of the gas. Solving for ρ, we find

$$\rho = \frac{p\mu}{kT}. \tag{8.30}$$

Substituting this back into the equation of hydrostatic equilibrium, we find

$$\frac{dp}{dz} = -\frac{p\mu}{kT}g. \tag{8.31}$$

To integrate this expression, we use the separation of variables technique:

$$\frac{dp}{p} = -\frac{\mu g}{kT}dz \tag{8.32}$$

$$\int \frac{dp}{p} = \int -\frac{\mu g}{kT}dz. \tag{8.33}$$

Here, we implicitly assume that T, g and μ are roughly constant with z—a reasonable, though not perfect, assumption. (A consideration of how "reasonable" this assumption is, and how the assumption affects $p(z)$ is provided as a homework exercise.) Then,

$$\ln p = -\frac{\mu g}{kT}z + C \tag{8.34}$$

$$p = Ce^{-\frac{\mu g}{kT}z}. \tag{8.35}$$

If we let $p(z = 0) = p_0$ then

$$p = p_0 e^{-\frac{\mu g}{kT}z}. \tag{8.36}$$

We have found that the pressure decreases exponentially with height. The fact that pressure decreases with height makes sense based on our stability argument—the pressure must be higher below the parcel than above the parcel, in order to produce a pressure gradient force pointed upwards. It also makes sense considering that the atmosphere near Earth's surface has more mass pushing on it from above than does the atmosphere far from Earth's surface.

The term $\frac{kT}{\mu g}$ has units of distance, and tells us how quickly the pressure decreases with height. Therefore, it is given its own name, the atmospheric *scale height*:

$$H = \frac{kT}{\mu g}. \tag{8.37}$$

Using this definition, we find:

$$p = p_0 e^{-\frac{z}{H}}. \tag{8.38}$$

We can also calculate the scale height for Earth's surface. With $T = 300$ K, $\mu = 29$ amu and $g = 9.8$ m s^{-2},

$$H = \frac{1.38 \times 10^{-23} \text{ m}^2 \text{ kg s}^{-2} \text{ K}^{-1} \times 300 \text{ K}}{29 \times 1.66 \times 10^{-27} \text{ kg} \times 9.8 \text{ m s}^{-2}} \tag{8.39}$$

$$\approx 8.8 \text{ km}. \tag{8.40}$$

So, if one goes up 8.8 km in Earth's atmosphere, the pressure will drop by a factor of e (≈ 2.7). It is this fact that makes the air thinner at the tops of mountains, and requires that mountain climbers sometimes use supplemental oxygen. In fact, the top of Mt. Everest is at a height of about 8.8 km, so the atmospheric pressure is about $1/2.7 \approx 37\%$ of the pressure at Earth's surface. Note that people sometimes say that there is less oxygen at the top of Mt. Everest. This is technically true—there is less oxygen... but also less of every other molecule as well. The oxygen level *as a percentage* of the total atmospheric composition is not appreciably different at the tops of mountains.

8.2.2 Convection and the Adiabatic Lapse Rate

In this section, we'll derive an expression for how the temperature changes with height near a planet's surface, and simultaneously address the question: why is it colder on the tops of mountains, and hotter in the bottom of Death Valley?

A planet's surface is heated primarily by sunlight. If the atmosphere is mostly transparent to visible wavelength radiation (as is true for the Earth, but not Venus) the surface heats up, making the bottom of the atmosphere warmer than the top. As the air warms, it expands, and becomes lighter than the surrounding air. It thus becomes buoyant relative to the surrounding atmosphere and rises. The rising of the warm air is balanced by a downward flow of cool air. This physical redistribution of heat is known as *convection*.

In the previous section, we learned that an atmosphere's overall pressure decreases with height. Therefore, as a parcel of air on a planet's surface rises, it finds itself surrounded by lower pressure air. It therefore expands, performing positive work on the environment. The environment, in turn, performs negative work on the parcel. Since work is proportional to change in internal energy ($W = \Delta U$) if no heat is exchanged, the parcel's internal energy, and temperature, must decrease as it rises. We can find the magnitude of this change in temperature on a hypothetical parcel of air as shown in Figure 8.6, experiencing a change in pressure Δp. Remember that work is the product of force and displacement, where the force exerted on the parcel is $\Delta p A$, and we can imagine displacing the parcel by Δz. Then:

$$W = \Delta U \tag{8.41}$$

$$\Delta p A \Delta z = m C_p \Delta T \tag{8.42}$$

$$\Delta p A \Delta z = \rho A \Delta z C_p \Delta T \tag{8.43}$$

$$\frac{1}{\rho} \Delta p = C_p \Delta T \tag{8.44}$$

where C_p is the heat capacity at constant pressure.

If we consider an infinitesimally small parcel, then we find:

$$\frac{1}{\rho} dp = C_p \, dT. \tag{8.45}$$

As expected, as the parcel rises (and pressure decreases), the parcel's temperature also decreases. When no heat is exchanged between the parcel and the environment, as we assumed here, this is known as an *adiabatic* process.

If we combine Equation (8.45) with the equation of hydrostatic equilibrium, we can find how temperature changes with height:

$$C_p \, dT = -g \, dz \tag{8.46}$$

$$\frac{dT}{dz} = -\frac{g}{C_p} \tag{8.47}$$

$$\Gamma = -\frac{g}{C_p}. \tag{8.48}$$

This expression is known as the *dry adiabatic lapse rate*, and is given the symbol Γ. (It is called the *dry* adiabatic lapse rate, because it does not account for any energy released due to condensation of water molecules.) This expression explains why the tops of mountains are cooler, and the bottoms of valleys are warmer.

For the Earth,

$$\Gamma \approx -\frac{9.8 \text{ m s}^{-2}}{10^3 \text{ J kg}^{-1} \text{ K}^{-1}} \tag{8.49}$$

$$\approx -9.8 \text{ K km}^{-1}. \tag{8.50}$$

As examples, the expected temperature change from sea level to the top of Mt. Washington in New Hampshire is ≈ -19 K ($\approx -34°$F), while the temperature change from the bottom to the top of the Grand Canyon is about ≈ -18 K ($-32°$F). Interestingly, if we try to apply this to Mt. Everest, we find a change of about -87 K ($-157°$F) from sea level to the peak, while actual temperatures at the summit only get as cold as about $-30°$F. It turns out that this discrepancy arises in part from the fact that Mt. Everest sits on top of the very large Tibetan plateau, whose surface directly absorbs stellar radiation. Mt. Everest is therefore much warmer than it would be if its base were at sea level.

We made several assumptions during this calculation. First, we assumed that the atmosphere was dry. This is correct when humidity is low enough that no water condenses; if water condenses, it releases heat that reduces the lapse rate. Second, we assumed that the heating came only from the Earth's surface, and that the atmosphere was convecting. Both of these are true only near the planet's surface, in a region known as the *troposphere*.

Figure 8.7 shows the temperature profiles for the terrestrial planets. The troposphere represents the lowest regions of these plots, where the atmosphere is convecting, and the temperature profile is determined by the adiabatic lapse rate. Higher in a planet's atmosphere, in a region known as the thermosphere, molecules directly absorb high energy radiation from the Sun. Since the heating comes from above in this case, the top of the thermosphere is warmer than the bottom. For the Earth, there are an additional two layers, known as the stratosphere and mesosphere, which form because of Earth's ozone layer, as shown in Figure 8.8. The ozone (O_3) molecules absorb radiation from the Sun, creating a feature known as a temperature inversion, where the temperature gradient with height reverses. Note that in atmospheric layers where temperature increase with height (thermospheres, and Earth's stratosphere), warmer air (which tends to rise) lies on top of cooler air (which tends to sink), so these layers do not undergo convection; we say that they are stably stratified.

8.3 The Greenhouse Effect

In this section, we'll discuss how the greenhouse effect can heat a planetary surface.

8.3.1 Greenhouse Gases and Surface Temperature

In Section 2.3 we derived the equilibrium temperature for planets by assuming that the energy absorbed by the planet equals the energy emitted by the planet. However, when we compare equilibrium temperatures to true temperatures, we find that terrestrial planets with thick atmospheres can have surface temperatures higher than

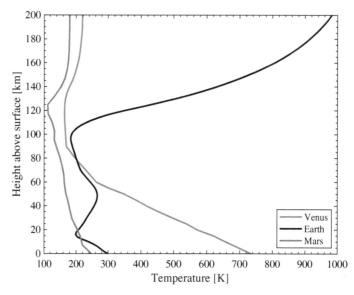

Figure 8.7. Temperature profiles for the terrestrial planets. In the convecting troposphere near the surface, temperature decreases with height following the adiabatic lapse rate. In the upper thermosphere, the temperature increases with height as molecules directly absorb radiation from the Sun. Only Earth has another reversal due to its Ozone layer. (Data for Venus from Hedin et al. 1983; Seiff et al. 1985, for Mars from the Martian Climate Database—Forget et al. 1999; Millour et al. 2018—and for Earth from the Community Coordinated Modeling Center NRLMSISE-00 Model—Labitzke et al. 1985; Hedin 1991; Picone et al. 2002.)

their equilibrium temperatures. For example, for Venus, the average surface temperature is 737 K (NASA 2020), while its equilibrium temperature (using Equation (2.17), assuming an albedo of 0 and emissivity of 1) would be only 327 K. How can this be? Is the planet therefore not in equilibrium?

In fact, the planet is still in equilibrium, but at a higher level, in the planet's atmosphere—not at its surface. This concept can be well understood by considering a simplified 1-layer atmosphere model, as shown in Figure 8.9, in which the atmosphere lets visible light through, but absorbs and reemits all infrared light. For this model, the energy inputs and outputs must balance both for the atmospheric layer and for the surface. At the top of the atmosphere, $E_{in} = E_{out,atm}$—the planet is, indeed, in equilibrium. But at the surface, there are now two inputs—the energy from the Sun, as well as the energy re-emitted by the atmosphere. Thus, $E_{in} + E_{out,atm} = E_{out,surf}$. These two inputs require that the output from the planet's surface, and therefore its temperature, be higher than that from the equilibrium temperature case. An important assumption in this model is that the atmosphere must be able to absorb radiation from the planet's surface, so that it heats up, but still let the Sun's visible-wavelength light reach its surface. An atmosphere has this property if it includes so-called *greenhouse gases*, which include H_2O, CO_2, CH_4, and O_3—molecules that readily absorb infrared radiation from a planet's surface, but are mostly transparent to visible light. The heating of a planet's surface through this effect is known (for historical reasons) as the *greenhouse effect*.

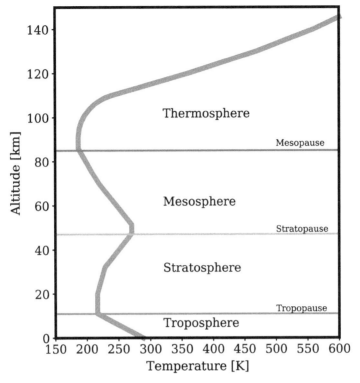

Figure 8.8. Temperature profile for the Earth, from the 1976 US Standard Atmosphere Model (COESA 1976). The so-called troposphere, stratosphere, mesosphere and thermosphere are defined by reversals in the temperature profile. The tropopause, stratopause and mesopause mark the boundaries between these regions.

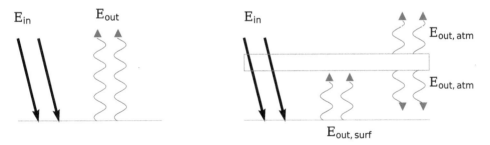

Figure 8.9. Simplified models showing an atmosphere-free planet (left), and a planet with a one-layer atmosphere (right). Black arrows represent visible wavelength radiation from the Sun, while red squiggly arrows represent infrared radiation from Earth's surface or atmosphere.

Why is this known as the greenhouse effect? Glass also has the property that it lets through visible wavelength radiation, but absorbs infrared radiation. It's known that greenhouses are warmer than their surroundings, so it was once thought that it was this property of greenhouses that caused them to heat up. However, analyses of the detailed energy balance have subsequently shown that the greenhouse effect on

Table 8.1. CO_2 Contents (Volume Percentages From Baines et al. 2013) and Total Masses (NASA 2020) for the Terrestrial Planets With Thick Atmospheres

Planet	CO_2 Volume Fraction	M_{atm} [kg]
Earth	3.9×10^{-4}	5.1×10^{18}
Venus	0.965	4.8×10^{20}
Mars	0.9597	2.5×10^{16}

actual greenhouses may be small. In fact, a primary reason actual greenhouses are warm is that the physical structure of the greenhouse inhibits convection, trapping the warm air heated by the Earth's surface.

8.3.2 CO_2 Content and Surface Temperatures of the Terrestrial Planets

One of the most important greenhouse gases, due to its ability to absorb infrared radiation, as well as its relatively high abundance in planetary atmospheres, is CO_2. Table 8.1 shows the CO_2 contents for the terrestrial planets, as well as their total masses. Both CO_2 content and overall mass contribute to the size of the greenhouse effect.[2] Note that Venus has both a high percentage of CO_2 in its atmosphere, as well as a high total atmospheric mass. Venus has the strongest greenhouse effect among the terrestrial planets. The Earth has only a small percentage of CO_2 in its atmosphere, but a massive enough atmosphere that it has a non-negligible greenhouse effect. While the Earth's equilibrium temperature is about 255 K (assuming an albedo of 0.3), its mean surface temperature is about 288 K. Mars has a high percentage of CO_2 in its atmosphere, but the atmosphere's overall mass is quite small, so the planet has a negligible greenhouse effect. Its equilibrium temperature is essentially the same as its mean surface temperature—about 210 K.

8.3.3 The Greenhouse Effect on Earth: Global Warming

CO_2 molecules are produced by the burning of Hydrocarbon-based fossil fuels. During the burning process, C–H bonds are broken, and the freed carbon molecules combine with the abundant Oxygen in Earth's atmosphere to form CO_2. Since the 1970s, the mass of CO_2 in Earth's atmosphere has increased by about 25% (see Figure 8.10). Since CO_2 is a greenhouse gas, this increase in CO_2 will increase (and has increased) the greenhouse effect on Earth. This phenomenon is known as *global warming, or climate change.*

Although the politics of trying to reduce fossil fuel use provides incentive for people to spread confusion about this topic, the relatively simple physics behind the greenhouse effect makes a clear prediction that increased CO_2 in the atmosphere will

---

[2] Note that the relationship between the amount of greenhouse effect and these two variables is nuanced; one cannot simply "count" the number of CO_2 molecules to estimate the size of the effect.

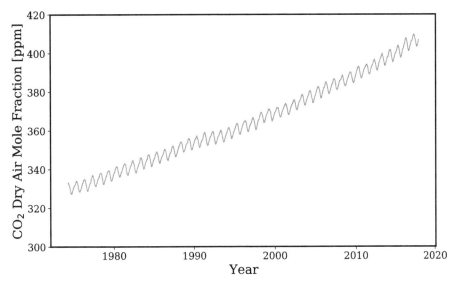

Figure 8.10. Earth's monthly average CO_2 content in parts per million (ppm) as measured on Mauna Loa, HI (NOAA 2018). (CO_2 dry air mole fraction means the number of CO_2 molecules as compared to the total number of atoms+molecules, with the exclusion of water vapor. Water vapor is excluded because it's variable due to its condensability.) Yearly variations are caused by seasonal changes in vegetation. The longer-scale upwards trend is anthropogenic (human-caused).

lead to increased temperatures on the surface. However, it is considerably more difficult to determine exactly how much temperature change is induced by a given increase in CO_2, and to make predictions about how future fossil fuel use will affect the global climate. These questions form the basis of much of climate change research. A primary strategy for tackling these questions is to create and analyze computer-based global climate models.

8.4 Atmospheric Dynamics

In this section, we'll briefly touch on the field of atmospheric dynamics, developing an understanding of basic wind patterns on planets, and large-scale climate on the Earth.

8.4.1 Hadley Circulation and Angular Momentum

Figure 8.11 shows a first-order picture for atmospheric circulation. Due to different sunlight angles, parts of the planet (labeled H, for Hot) are warmer than others. For planets with low obliquity (the angle between the planet's polar and orbital angular momentum vectors) like the Earth, the warmer regions lie near the equator, while the cooler regions (labeled C, for cold) are near the poles. The atmosphere attempts to redistribute this heat, allowing the hot air to rise and move poleward, while the cool air sinks, and moves toward the equator. This type of atmospheric circulation is known as *Hadley circulation*, after George Hadley, who proposed this theory.

Figure 8.11. Basic diagram showing expected circulation on a planet heated by sunlight. The hot (H) and cold (C) regions are determined by the angle of the Sun's radiation. The hot air rises and is redistributed toward the cold region. Colder air sinks and is redistributed toward the hot region. Not to scale.

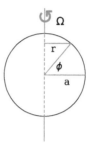

Figure 8.12. At the equator, the atmosphere is a distance a from the polar axis, while at other latitudes, ϕ, it is at some smaller distance r, equal to $a \cos \phi$ (for a perfect sphere). Because the distance to the rotation axis changes as air moves from equator to pole, the rotation speed of the air must change to conserve angular momentum.

This circulation, when subject to the principle of conservation of angular momentum, produces planetary-scale wind patterns. Consider the diagram in Figure 8.12. A particle at the planet's equator has angular momentum $L = m\omega_{eq}a^2$ or *specific angular momentum* (angular momentum per unit mass) of $A = \omega_{eq}a^2$, where ω_{eq} is the particle's angular velocity at the equator and a is the equatorial radius. As the particle travels toward the pole, its specific angular momentum at any given location is given by $A = \omega r^2$, where r is the distance to the polar axis. Since angular momentum is conserved, $\omega_{eq}a^2 = \omega r^2$; as the particle travels toward the pole, r decreases, and ω must therefore increase. Since the surface itself is undergoing solid body rotation (i.e., ω_{surf} is always constant), this process produces so-called *westerly* winds, which travel *from* west to east as seen by an observer on the ground. Similarly, as air at the pole moves toward the equator, its ω must decrease as r increases to conserve angular momentum. As viewed by an observer on the ground, this produces so-called *easterly* trade winds, which travel from east to west. This idea is diagrammed in Figure 8.13.

However, as we will show, the principle of conservation of angular momentum also makes it difficult for planets to maintain single-cell Hadley circulation. The specific angular momentum of a particle of air can be separated into a term due to Earth's rotation, and a term representing the horizontal wind speed relative to Earth's surface, u,

$$A = \Omega r^2 + ur \qquad (8.51)$$

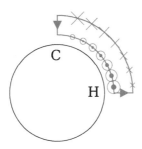

Figure 8.13. Diagram showing changes in wind speeds, assuming conservation of angular momentum, as air travels toward and away from the pole, with dot-filled circles and x's representing the heads and tails, respectively, of wind vectors.

where r is the distance to the polar axis and Ω is the angular speed of the planet's rotation. Figure 8.12 shows that, for a spherical planet, the distance from the surface to the polar axis is given by $r = a \cos \phi$, where ϕ is latitude. Therefore,

$$A = \Omega a^2 \cos^2 \phi + ua \cos \phi. \tag{8.52}$$

If we let the air at the equator have $u = 0$ (which is roughly true for the Earth), and since we know that at the equator $\cos(\phi) = 1$, the specific angular momentum at the equator is given by $A_0 = \Omega a^2$. Conservation of angular momentum then says:

$$\Omega a^2 = \Omega a^2 \cos^2 \phi + ua \cos \phi \tag{8.53}$$

or if we solve for u:

$$ua \cos \phi = \Omega a^2 (1 - \cos^2 \phi) \tag{8.54}$$

$$u = \frac{\Omega a^2 \sin^2 \phi}{a \cos \phi} \tag{8.55}$$

$$u = \Omega a \frac{\sin^2 \phi}{\cos \phi}. \tag{8.56}$$

Table 8.2 shows calculated wind speeds, u, given this equation, for locations on Earth. Category 1 hurricanes have wind speeds of 74–95 mph, and the sound speed at the surface of the Earth is about 760 mph, so clearly these winds are not representative of realistic conditions on Earth. In contrast, on Earth, the circulation pattern breaks up into a series of 3 cells per hemisphere (6 total), as shown in Figure 8.14. This pattern of circulation has wide-reaching implications for Earth's climate. At the Earth's equator, rising air cools, creating a warm, wet climate. In contrast, between the first (Hadley) and second (Ferrel) cells, where the air sinks, the air is depleted in water, and as it sinks, it warms, further increasing its ability to maintain water in the vapor phase. Thus, rain is extremely rare. This location, near 30° latitude, corresponds with the so-called desert belt on Earth. The Sonoran, Mojave, Atacama, Arabian, Syrian, and Sahara deserts (among others) are located

Table 8.2. Calculated Horizontal Wind Speeds using Equation (8.56)

Latitude (ϕ)[°]	u [m s^{-1}]	u [mph]
30°	130	290
45°	324	724
60°	687	1500

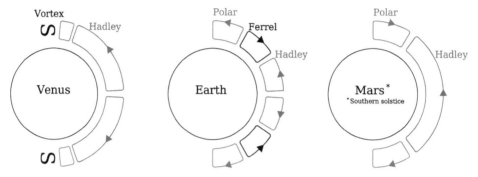

Figure 8.14. Atmospheric circulation cells on Venus, Earth, and Mars. Slowly-rotating Venus has large Hadley circulation cells, S-shaped polar vortices, and "collars" of cold air in-between. Faster-rotating Earth has a circulation pattern in which the equator-to-pole circulation is separated into three cells: a Hadley cell, mid-latitude Ferrel cell and high-latitude polar cell. Mars has a circulation pattern that depends on season. Near equinox, Mars supports two Hadley cells, like on Earth. However, at other times, Mars develops a single Hadley cell transporting from the summer to winter hemisphere, and two polar cells. Karatekin et al. (2011).

close to this latitude. Between the second (Ferrel) and third (Polar) cells, where the air rises again, relatively warm surface air rises and cools, causing precipitation. This location is associated with temperate rainforests, such as that of the US Pacific northwest.

8.4.2 Geostrophic Balance

If you look carefully at Figure 8.13, you can see that, while the air traveling toward the pole gets deflected eastward, and the air traveling toward the equator gets deflected westward, in both cases, the air is deflected *to the right*. This tendency for things to get deflected to the right is exactly the same as the Coriolis effect we discussed in Section 4.1.1. Talk yourself through the process in the southern hemisphere, and you'll find that the opposite happens there—the air will get deflected to the left. As we'll describe, in addition to creating the large-scale wind patterns described in the previous section, the Coriolis force plays a role in small-scale wind patterns as well.

In Section 8.2.1 we learned about the pressure gradient force, which results from differences in pressure, and tends to move things from high pressure to low pressure. Local weather creates high and low pressure regions, and when this occurs, the air feels a force pushing it from high pressure regions to low. Like a popped balloon, we might naively expect the air to rush out from high pressure zones, or rush into low pressure zones, as shown in Figure 8.15. However, on a rotating planet, the air also feels a Coriolis force, deflecting it toward the right. The air can get deflected until it is traveling parallel to the pressure gradient, at which point the pressure gradient force is opposite the direction of the Coriolis force, and the two can balance out. This balance is known as *geostrophic balance*, and tends to cause winds to travel, somewhat non-intuitively, parallel to, rather than across, pressure gradients. (In reality, geostrophic balance is an approximation, as it does not account for complications like varying pressure gradients, and frictional forces.)

As you might deduce from Figure 8.15, geostrophic balance is the reason behind the wind patterns we experience in large storms on Earth, like hurricanes. Both low and high pressure zones can produce spiral wind patterns, although they flow in opposite directions called either cyclonic or anticyclonic, respectively. However, the high pressure cores push some air away from the core, causing a downdraft that warms as it decreases in altitude. Much like at the bottoms of Hadley cells, the warm air is much less likely to produce precipitation, as it can easily accommodate a lot of moisture in the gas phase. In contrast, low pressure zones move air toward their center and upwards, and as this air rises and cools, it condenses to form clouds and precipitation. Therefore, low pressure zones are the ones that produce large storms, like hurricanes. Recall that the Coriolis force deflects toward the *right* in the Northern hemisphere, but toward the *left* in the Southern hemisphere. Therefore, Northern hemisphere hurricanes flow counterclockwise, while the Southern hemisphere equivalent—typhoons—flow clockwise.

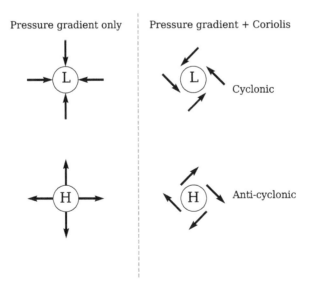

Figure 8.15. Northern hemisphere patterns around a low pressure zone (marked "L") and high pressure zone (marked "H") without Coriolis forces (left) and with large Coriolis forces (right).

8.4.3 Wind Patterns on Solar System Planets

The atmospheres of the planets in the solar system each have features analogous to those on Earth, although the details are slightly different for each planet. The circulation patterns of the terrestrial planets with large atmospheres (Venus, Earth, and Mars) are compared in Figure 8.14. Venus rotates very slowly (with a period of 243 Earth days) and so the angular momentum at its equator is not that high. Therefore, Venus does actually maintain a single-cell Hadley-style circulation (e.g., Karatekin et al. 2011). Venus also supports S-shaped polar vortices, and cold "collars" northward of the Hadley cell. Mars has a similar circulation pattern to Earth when it is close to the equinoxes (when the planet's pole is pointed neither toward nor away from the Sun). However, at other times upper latitudes receive more direct radiation than the equator, and the atmosphere develops a 3 cell circulation pattern, with one cell carrying warm air from the summer hemisphere to the winter hemisphere (e.g., Leovy 1969; Forget et al. 1999).

The giant planets all have multi-cell circulation patterns with high wind speeds. For example, Figure 8.16 shows the horizontal wind speeds for Jupiter and Saturn, measured by tracking the motion of features in the planets' atmospheres using spacecraft imaging. The cells' upwelling and downwelling are the cause of a dramatic series of light and dark stripes on Jupiter, known as zones and belts, respectively. The zones and belts are apparent when viewing Jupiter with even a modest-sized telescope. Jupiter's atmospheric structure causes a series of different clouds to condense as a function of altitude. Where upwelling occurs, we see the higher altitude, light-colored clouds; where downwelling occurs, we see a deeper layer composed of darker clouds. Figure 8.16 shows how the stripes are associated

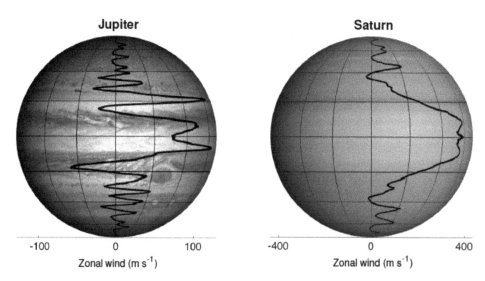

Figure 8.16. Horizontal wind speeds (U) on Jupiter and Saturn, as a function of latitude. Reprinted by permission from SpringerNature: Springer, Kaspi et al. (2020), © 2020.

with *changes* in horizontal wind speed—i.e., locations where the atmospheric cells are upwelling or downwelling.

As on Earth, other planets in the solar system host cyclonic and anticyclonic wind cells. However, on a giant planet, where friction is minimized by the lack of a solid surface, these storms can be very long-lived. Jupiter's Great Red Spot is an anticyclonic (high pressure) storm that has existed for at least hundreds of years (based on recorded observations). Its diameter is even larger than the diameter of the Earth, and wind speeds reach hundreds of miles per hour.

8.5 Important Terms

- exosphere;
- escape velocity/escape speed;
- the Maxwell–Boltzmann distribution;
- pressure gradient force;
- the equation of hydrostatic equilibrium;
- scale height;
- convection;
- adiabatic;
- dry adiabatic lapse rate;
- troposphere;
- greenhouse gas;
- the greenhouse effect;
- global warming/climate change;
- Hadley circulation;
- specific angular momentum;
- easterlies/westerlies;
- geostrophic balance.

8.6 Chapter 8 Homework Questions

1. **Scale height:**

    When we derived scale height, $H = \frac{kT}{\mu g}$, in Section 8.2.1, we assumed that T, μ and g were constant, and found that $H \approx 8.8$ km. In this problem, we'll test the validity of these assumptions by considering how H would be different if you recalculated it using values of these parameters at the mesopause. The height and temperature of the mesopause can be estimated from Figure 8.8.

    (a) First, recalculate H to at least 3 significant digits using default parameters.

    (b) Recalculate H by using the value of g at the mesopause, keeping other parameters constant. Discuss the validity of assuming g is constant.

(c) Recalculate H by using the value of T at the mesopause, keeping other parameters constant. Discuss the validity of assuming T is constant.

2. **Fossil fuels and the greenhouse effect:**

In this question, we'll perform some rough calculations mirroring those performed by climate researchers.

(a) Vehicles using fossil fuels (i.e., gasoline) are a primary contributor to climate change. Using reasonable assumptions, estimate the total numbers of gallons of gasoline used by all drivers on planet Earth over the course of a year.

(b) Assuming for simplicity that gasoline is made entirely of Octane (C_8H_{18}), and that a gallon of gasoline weighs about 3 kg, estimate the number of C atoms per gallon of gasoline. Assuming each and every C atom combines with O_2 to make CO_2, this will also be the total number of CO_2 molecules produced per gallon of gasoline burned.

(c) Based on your answers to the last two questions, and assuming fuel usage has remained constant, estimate the total number of CO_2 molecules produced from 1970–2020.

(d) Using the total mass of Earth's atmosphere (see Table 8.1), and the fact that the atmosphere is primarily composed of N_2, estimate the total number of molecules in Earth's atmosphere.

(e) Using your answers to the last two questions, estimate the increase in fraction of CO_2 in parts per million over the years 1970–2020.

(f) Compare your answer to the increase in CO_2 fraction shown in Figure 8.10. According to your calculation, to what extent is the use of gasoline by drivers contributing to CO_2 production? What other factors might you take into account to better match Figure 8.10?

(g) Try repeating this calculation for another contributor to CO_2 emissions. What did you learn?

References

Baines, K. H., Atreya, S. K., Bullock, M. A., et al. 2013, Comparative Climatology of Terrestrial Planets (Tucson, AZ: Univ. of Arizona Press), 137

COESA 1976, U.S. Standard Atmosphere (Washington, DC: U.S. Government Printing Office)

Forget, F., Hourdin, F., Fournier, R., et al. 1999, JGR, 104, 24155

Hedin, A. E., Niemann, H. B., Kasprzak, W. T., et al. 1983, JGR, 88, 73

Hedin, A. E. 1991, JGR, 96, 1159

Jansen, T., Scharf, C., Way, M., et al. 2019, ApJ, 875, 79

Karatekin, Ö., de Viron, O., Lambert, S., et al. 2011, P&SS, 59, 923

Kaspi, Y., Galanti, E., Showman, A. P., et al. 2020, SSRv, 216, 84

Labitzke, K., Barnett, J. J., & Edwards, B. 1985, Middle Atmosphere Program. Handbook for MAP, Vol. 16, Atmospheric Structure and Its Variation in the Region 20 to 120 Km, Draft of a New Reference Middle Atmosphere (Chestnut Hill, MA: SCOSTEP)

Leovy, C. B. 1969, ApOp, 8, 1279

McClintock, W. E., Cassidy, T. A., Merkel, A. W., et al. 2018, in Mercury: The View after MESSENGER (Cambridge Planetary Science), ed. S. Solomon, L. Nittler, & B. Anderson (Cambridge: Cambridge Univ. Press), 371

Millour, E., Forget, F., Spiga, A., et al. 2018, in Mars Science Workshop, From Mars Express to ExoMars, 68

NASA/Goddard Space Flight Center 2020, Space Science Data Coordinated Archive: Planetary Fact Sheet, https://nssdc.gsfc.nasa.gov/planetary/factsheet/

NOAA ESRL Global Monitoring Division 2018, Updated annually. Atmospheric Carbon Dioxide Dry Air Mole Fractions from Quasi-continuous Measurements at Mauna Loa, Hawaii. Compiled by K. W., Thoning, D. R., Kitzis, & A. Crotwell, National Oceanic and Atmospheric Administration (NOAA), Earth System Research Laboratory (ESRL), Global Monitoring Division (GMD): Boulder, CO, Version 2018-1

Picone, J. M., Hedin, A. E., Drob, D. P., et al. 2002, JGRA, 107, 1468

Porco, C. C., West, R. A., McEwen, A., et al. 2003, Sci, 299, 1541

Sanchez-Lavega, A., Rojas, J. F., & Sada, P. V. 2000, Icar, 147, 405

Seiff, A., Schofield, J. T., Kliore, A. J., et al. 1985, AdSpR, 5, 3

University of Nebraska, Lincoln 2020, The Nebraska Astronomy Applet Project 2020, Atmospheric Retention Lab, https://astro.unl.edu/naap/atmosphere/atmosphere.html

Introductory Notes on Planetary Science
The solar system, exoplanets and planet formation
Colette Salyk and Kevin Lewis

Chapter 9

Planet Formation

9.1 Star and Disk Formation

Why do stars and planets form at all? Is their formation linked? These are the questions we'll address in this section.

9.1.1 Gravity–Pressure Balance and the Jeans Mass

Consider a gas cloud sitting somewhere in the Galaxy. (Why do we call it a gas cloud? Since the mass ratio of gas to solids in the universe is around 100, we can often safely ignore the presence of solids in many situations. In addition, there is so much hydrogen in the universe, assuming the cloud is made purely of hydrogen is also not a bad assumption in many cases.) In order to turn this gas cloud into a star, the force of gravity must pull the cloud together. However, the gas also has pressure, due to the constant motions of the molecules, and this pressure resists the gravitational collapse. (Magnetic forces may also play a role in resisting collapse, but we won't get into that here.) If we know some basic properties of the gas, we can estimate the minimum cloud mass for which the force of gravity will overcome the pressure, and cause the cloud to collapse. This mass is known as the *Jeans mass*, named after the British astronomer Sir James Jeans.

In order to estimate the Jeans mass, we assume that the gravitational potential energy of the cloud (E_{grav}) is similar to the total kinetic energy associated with the random motions of the gas particles (E_{gas}). In fact, "similar" specifically means that $E_{grav} = 2 \times E_{gas}$—a relation known as the Virial theorem (that we won't derive here). When this is true, the cloud is just on the verge of collapse. If the mass gets any higher, the cloud collapses.

The gravitational potential energy of a uniform-density sphere of mass M and radius R is

$$E_{grav} = \frac{3}{5}\frac{GM^2}{R} \tag{9.1}$$

—an equation we derived in Section 6.3.1.

doi:10.1088/2514-3433/abb198ch9   © IOP Publishing Ltd 2020

The kinetic energy of N atoms of a monoatomic ideal gas at a temperature T is:

$$E_{gas} = \frac{3}{2}NkT \tag{9.2}$$

where k is the Boltzmann constant. If we know the average mass per atom, μ, we can express this in terms of the total mass of the cloud, M:

$$E_{gas} = \frac{3}{2}\frac{M}{\mu}kT. \tag{9.3}$$

We can use the Virial theorem to relate the gravitational and kinetic energies and then solve for the mass at which the cloud is just on the verge of collapse. This is the Jeans mass, which we'll call M_J. By expressing the Jeans mass in terms of the temperature and density of the gas, and its average atomic mass, we allow the Jeans mass to be estimated using cloud properties that can actually be measured via observations.

First, let's express R_J as a function of M_J to remove the dependence on R_J:

$$M_J = \frac{4}{3}\pi R_J^3 \rho \tag{9.4}$$

$$R_J = \left(\frac{3M_J}{4\pi\rho}\right)^{1/3} \tag{9.5}$$

and then relate the gravitational and kinetic energies:

$$E_{grav} = 2E_{gas} \tag{9.6}$$

$$\frac{3}{5}\frac{GM_J^2}{R_J} = 2 \times \frac{3}{2}\frac{M_J}{\mu}kT \tag{9.7}$$

$$\frac{3}{5}GM_J^{5/3}\left(\frac{4\pi\rho}{3}\right)^{1/3} = 3\frac{M_J}{\mu}kT \tag{9.8}$$

$$M_J^{2/3} = \frac{kT}{\mu}\frac{5}{G}\left(\frac{3}{4\pi\rho}\right)^{1/3} \tag{9.9}$$

$$M_J = \left(\frac{5kT}{\mu G}\right)^{3/2}\left(\frac{3}{4\pi\rho}\right)^{1/2}. \tag{9.10}$$

The Jeans mass tells us the minimum mass at which a gas cloud will collapse. Using this expression one can then infer what properties make it "easier" for a cloud to collapse—namely, lower temperatures, and/or higher densities.

If you compute the Jeans mass for typical cloud parameters (as you will try for homework), you'll find that it's much larger than the mass of the Sun, and therefore much larger than the masses of planets. Understanding why the Jeans mass is larger than typical stellar masses is an ongoing topic of research in star formation. For our purposes, we'll note that this calculation shows us that planets are unlikely to spontaneously collapse in space from a cloud of gas. Instead, as we'll show in the next section, planets form as a byproduct of star formation.

9.1.2 Angular Momentum and Disk Formation

So why do planets form? As we'll see in this section, planets exist due to the angular momentum associated with gas clouds as they collapse to form stars. If we approximate a gas cloud as a uniform sphere, its angular momentum, L, is given by:

$$L = I\omega \tag{9.11}$$

$$L = \frac{2}{5}MR^2\omega \tag{9.12}$$

where I is moment of inertia, ω is the angular velocity of the cloud, and M and R are the cloud's mass and radius. As the cloud collapses, it must conserve angular momentum, and therefore it must spin up as it gets smaller. Supposing, for example, that the cloud has some initial and final radii, R_i and R_f, respectively, we can see mathematically that:

$$L_i = L_f \tag{9.13}$$

$$\frac{2}{5}MR_i^2\omega_i = \frac{2}{5}MR_f^2\omega_f \tag{9.14}$$

$$\omega_f = \omega_i\left(\frac{R_i}{R_f}\right)^2. \tag{9.15}$$

Since $R_i > R_f$, the cloud will spin up as it collapses. This is the same physical principle that causes the water in your sink or bathtub to spin ever faster as it drains.

In addition, as the cloud collapses, the parts of the cloud close to the rotation axis feel a gravitational force but little to no centrifugal force; in contrast, the parts of the cloud far from the rotation axis feel a significant centrifugal force acting counter to the gravitational force. This idea is illustrated in Figure 9.1. The net forces cause the disk to flatten as it collapses, eventually forming a centrally condensed region (the protostar) and a surrounding disk. This process is analogous to what happens when a chef spins pizza dough. Although the dough begins in a clump, the spinning causes the dough to bulge outwards, creating a disk shape.

Another way to think about disk formation is that as the cloud collapses, conservation of angular momentum increases the rotation rate of the cloud so much that it simply cannot retain a spherical shape. The disk therefore serves as a

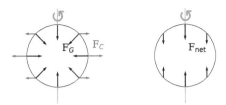

Figure 9.1. Simplified cartoon illustration of forces in a rotating, star-forming cloud. Left: Centrifugal forces, colored red and labeled F_C, always point perpendicular to the rotation axis. Gravitational forces, colored black and labeled F_G, point toward the center of the cloud. Right: Net forces push material into a disk-like shape.

reservoir to hold angular momentum so that angular momentum can be conserved while the central part of the cloud still collapses to form a star. Eventually, the disk will form a planetary system and the planets become the primary carrier of the initial angular momentum. Thus, planetary systems essentially form because of conservation of angular momentum. As a homework question, you'll calculate the angular momentum of a typical molecular cloud, and compare to the angular momentum of the Sun and the planets.

9.2 Protoplanetary Disk Properties

9.2.1 Bulk Properties: Composition and Mass

The disks that form due to cloud collapse and conservation of angular momentum are known as *circumstellar disks* or *protoplanetary disks*. For the solar system specifically, this disk is also referred to as the *solar nebula*. What are these disks made of? Since they form as a byproduct of star formation, they are made of the same ingredients as stars. Using the Sun as an example, if we take the solar abundances of the elements and apportion them into rocks (like olivine, or quartz) and gases (like H_2 and He), we'll find that the mass ratio of gas to rocks in the early solar system must have been about 100. (You'll derive a more precise value yourself as a homework question.) Astronomers call the small rocks dust, and this mass ratio of gas to rocks the *gas-to-dust ratio*. To summarize, protoplanetary disks are composed of gas and dust, in a ratio of about 100-to-1.

How massive are typical protoplanetary disks? One estimate of disk mass can be calculated using the solar system planets. However, we can't just add up the planetary masses, since the presence of terrestrial planets (for which the current gas-to-dust ratio is much less than 100) tells us that some of the original gas in the gas cloud must have been lost. For homework, you'll estimate the amount of missing gas, then add up the estimated "original" masses of the solar system planets (before any gas was lost) to estimate the original mass of the solar system (Hayashi 1981). This is known as the so-called *Minimum Mass Solar Nebula (MMSN)*, where minimum mass refers to the minimum disk mass required to form the planets of the solar system. As you will find for homework, this mass is approximately a few hundredths of a solar mass.

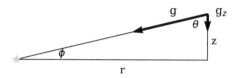

Figure 9.2. Disk geometry showing the directions of the gravitational acceleration, g, and its z component, g_z at some distance r from the star.

9.2.2 The Shape of Protoplanetary Disks: Scale Height and Disk Pressure Gradients

Because the protoplanetary disk is composed primarily of gas, it is actually a kind of atmosphere, and we can use the principles of planetary atmospheres to derive some properties of disks. In particular, the concept of hydrostatic equilibrium (balance between the gravitational force and pressure gradient force—see Equation (8.28)) determines the vertical structure of both planetary atmospheres and protoplanetary disks. For a planetary atmosphere the gravitational force acts anti-parallel to the vertical pressure gradient force. (In other words, the gravitational force points inwards, toward the center of the planet, while the pressure gradient force points outwards, away from the planet's surface.) However, for a protoplanetary disk, the gravitational force points toward the star,[1] and only a small component of this force points anti-parallel to the vertical pressure gradient force, as shown in Figure 9.2. In addition, the gravitational force has no vertical component at the midplane of the disk, but grows symmetrically as you get farther from the midplane in either direction. Finally, the magnitude of the gravitational force depends on the distance from the star, with the farthest reaches of the disk feeling the weakest gravitational force. Given these facts, we can expect that the highest pressure exists at the midplane of the disk, and drops off with height, and that the pressure is highest near the star and drops off as we get farther from the star.

Now that we have some intuitive understanding of what to expect, let's use the assumption of hydrostatic equilibrium to find an expression for $p(z)$—pressure as a function of height above the disk midplane, just like we did for planetary atmospheres. For a disk atmosphere,

$$\frac{dp}{dz} = -\rho g_z \tag{9.16}$$

where p is pressure, z is height above the disk midplane, ρ is the density of the gas, and g_z is the z component of the gravitational acceleration provided by the central star. To find g_z, we remember that the force of gravity from the star has a magnitude $F_G = \frac{GMm}{r^2}$, where M is the mass of the star, m is the mass of a particle, and r is the distance of the particle from the star, such that the gravitational acceleration, $g = \frac{GM}{r^2}$. In addition, we use the geometry shown in Figure 9.2 to show that the

---

[1] This assumes that the star is much more massive than the disk, such that the gravitational force from the star dominates over that from the disk itself.

z component of the acceleration is given by $g_z = g \cos \theta = g \sin \phi$. Since the disk is likely to be thin, we can assume $\phi \ll 1$, such that $\sin \phi \approx \tan \phi = \frac{z}{r}$. Therefore,

$$g_z = \frac{GM}{r^2} \frac{z}{r} \qquad (9.17)$$

$$= \frac{GM}{r^3} z. \qquad (9.18)$$

At this point, we can use our knowledge of planetary orbits to make a neat substitution into Equation (9.18). For circular motion caused by gravity, we know that:

$$\frac{GMm}{r^2} = \frac{mv^2}{r}, \qquad (9.19)$$

where M is the mass of the central body, m the mass of the orbiting body, v the speed of the orbiting body, and r the radius of the orbit. Then,

$$\frac{GM}{r} = v^2 \qquad (9.20)$$

$$\frac{GM}{r} = (\Omega r)^2 \qquad (9.21)$$

$$\Omega^2 = \frac{GM}{r^3} \qquad (9.22)$$

where Ω is the angular speed of the orbit. Therefore, we can rewrite g_z in terms of Ω as

$$g_z = \Omega^2 z \qquad (9.23)$$

and we are now ready to substitute this expression into the equation for hydrostatic equilibrium.

Recalling that the ideal gas law relates density (ρ), temperature (T) and pressure (p),

$$\rho = \frac{p\mu}{kT} \qquad (9.24)$$

where μ is the mean molecular mass, we find

$$\frac{dp}{dz} = -\frac{p\mu}{kT}\Omega^2 z. \qquad (9.25)$$

In order to find $p(z)$, we use separation of variables and integrate both sides:

$$\frac{dp}{p} = -\frac{\mu}{kT}\Omega^2 z \, dz \qquad (9.26)$$

$$\ln p = -\frac{\mu}{2kT}\Omega^2 z^2 + C \tag{9.27}$$

$$p = p_0 e^{-\frac{\mu}{2kT}\Omega^2 z^2} \tag{9.28}$$

where p_0 represents the pressure at $z = 0$ (i.e., at the disk midplane).

Just as with planetary atmospheres, we can make a substitution using the definition of the speed of sound:

$$c_s^2 = \frac{kT}{\mu}. \tag{9.29}$$

Then,

$$p = p_0 e^{-\frac{1}{2}\frac{\Omega^2}{c_s^2}z^2}. \tag{9.30}$$

Finally, we can define a scale height for the disk atmosphere to be $H \equiv \frac{c_s}{\Omega}$. Then,

$$p = p_0 e^{-\frac{1}{2}(\frac{z}{H})^2}. \tag{9.31}$$

Using this expression, we can see that the pressure drops off exponentially with height, as it does for a planetary atmosphere, but that the atmosphere drops off in both directions (above and below the midplane of the disk), as the pressure is symmetric with respect to height, z.

Take a moment to think about what H means—it's telling you how quickly the pressure drops off with height. For example, at a height of one scale height ($z = H$), the pressure will have decreased to $e^{-1/2} \approx 0.6$ of its midplane value. At a height of two scale heights ($z = 2H$), the pressure will have decreased to $e^{-2} \approx 0.14$ of its midplane value, and so on. The scale height therefore gives a representative thickness of the disk atmosphere.

What is the approximate size of the scale height H? Well, first, we must recognize that since H depends on T (via the sound speed equation) and Ω, it will depend on distance from the star. Let's consider 1 au and assume the disk atmosphere is dominated by molecular hydrogen, such that its average molecular mass, $\mu \approx 2 \times 1.66 \times 10^{-27}$ kg. At 1 au, the disk temperature may be near the temperature of Earth (≈ 300 K), so the sound speed, $c_s = \sqrt{kT/\mu} \approx 1000$ m s^{-1}. The Keplerian rotation speed is $\Omega = \frac{2\pi}{365 \times 24 \times 60 \times 60}$ s$^{-1} \approx 2 \times 10^{-7}$ s^{-1} and so the scale height, $H = \frac{c_s}{\Omega} \approx 5 \times 10^9$ m. If we express this in relation to the distance from the star, $\frac{H}{r} = \frac{5 \times 10^9 \text{ m}}{1.5 \times 10^{11} \text{ m}} \approx 1/30$. Therefore, we see that disks are much wider than they are tall; they are geometrically thin.

How does H vary as the distance from the star increases? To figure this out, we have to consider the dependence of both Ω and T (and therefore c_s) on distance from the star. Since $\Omega = \sqrt{\frac{GM_*}{r^3}}$, $\Omega \propto r^{-3/2}$. However, the dependence of T on r is not at all trivial to determine, as it depends on what the dust in the disk is made of (dark and small grains of dust will absorb more light overall), how they are distributed in the disk, how the disk is shaped (for example, a very flat disk will absorb less starlight than a very puffy disk), etc. Let's assume that $T \propto r^p$ where p is some unknown constant exponent that can be determined empirically via observations. Then, $c_s \propto T^{1/2} \propto r^{p/2}$ and $H = \frac{c_s}{\Omega} \propto \frac{r^{p/2}}{r^{-3/2}} \propto r^{\frac{p+3}{2}}$. Values for p are measured to be in the range of -0.5 to -0.6 (e.g., Andrews et al. 2009), so let's try substituting in $p = -0.5$ (which is, in fact, how planetary equilibrium temperature depends on distance from the star—see Equation (2.15)). Then, $H \propto r^{1.25}$. Thus, H increases with r, but in a fashion that is somewhere between linear and squared. Since H represents the thickness of the disk, this function tells us how the disk is shaped. As shown in Figure 9.3, such a disk is actually *flared*.

9.2.3 Disks Are Dusty: Optical Depth

So far, we've been treating the protoplanetary disk as an atmosphere—implicitly assuming it's primarily made of gas. However, while the gas dominates the mass of the disk, it is transparent to light at most wavelengths. Therefore, just as we get a bright sunny day as long as there are no clouds in the way, a protoplanetary disk would be nearly transparent if it didn't also include dust.

Exactly how dusty (or opaque) the disk is depends on the size of the dust grains. For example, once the dust has collected into planets, the planets occupy only a small amount of space in the disk, and light easily makes it through the disk. However, if we were to split up those planets into tiny particles, these would occupy a much larger portion of space, blocking a lot of light, as does smoke from a fire. Studies of protoplanetary disks show that they actually contain large amounts of approximately 1 µm-sized grains. Suppose we were to eventually construct an Earth-like planet from such grains—how many would we need?

The number of 1 µm-sized grains needed to build an Earth would be:

$$N_{\text{grains}} = \frac{M_{\oplus}}{M_{\text{grain}}} \tag{9.32}$$

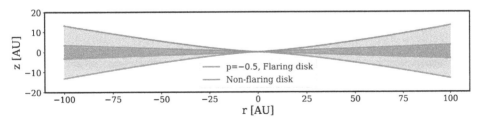

Figure 9.3. Scale height, H, versus distance from the star, r, for two representative disks, demonstrating possible protoplanetary disk shapes. The curved plot shows the shape of the disk atmosphere for the case where $T \propto r^{-0.5}$ ($H \propto r^{1.25}$), while the straight line shows a non-flaring disk with $H/r = 30$.

$$= \frac{4/3\pi R_{\oplus}^3 \, \rho_{\oplus}}{4/3\pi R_{\text{grain}}^3 \, \rho_{\text{grain}}}. \tag{9.33}$$

Assuming the grains have approximately the same average density as Earth, we find:

$$N_{\text{grains}} = \frac{R_{\oplus}^3}{R_{\text{grain}}^3} \tag{9.34}$$

$$\approx \left(\frac{6400 \times 10^3 \text{ m}}{10^{-6} \text{ m}} \right)^3 \tag{9.35}$$

$$\approx 10^{38}. \tag{9.36}$$

Wow! By this calculation, it would have taken 10^{38} grains to form an Earth-like planet. And, we might imagine that this many dust grains could begin to make the disk opaque. But in order to know for sure, we'll need to quantify the opaqueness of the disk. This is measured with a quantity called *optical depth*.

Optical depth, represented with the symbol τ, is a way to represent the percentage of light that will make it through a column of material. More precisely, if there is an incoming light source with flux F_0, then the fraction of light exiting the other side is given by $\frac{F}{F_0} = e^{-\tau}$. The larger the optical depth is, the less light gets through to the other side, and for *small* values of τ, the fraction of light making it through the column can be approximated by $1 - \tau$. (You can derive this fact for yourself by taking a Taylor expansion of $F(\tau)$ for $\tau \ll 1$. It turns out the first two terms in the expansion are $e^{-\tau} \approx 1 - \tau$.) For example, if $\tau = 0.1$, 90% of the light makes it through, while if $\tau = 0.2$, 82% of the light makes it through. For large values of τ, very little light makes it through; for example, for $\tau = 10$, only 0.005% of the light makes it through. For situations where the wavelength of light is smaller than the dust grain size, you can also think of τ conceptually as the percentage of cross-sectional area of the column that contains grains, as illustrated in Figure 9.4. When τ is small, the cross-sectional area of the grains is smaller than the cross-sectional area of the column, and light is likely to pass through the column without hitting a dust grain; the reverse is true if τ is large. When $\tau = 1$, the cross-sectional area of the dust grains is similar to that of the column. However, since some dust grains are likely to overlap each other in space, some light will still make it through the column.

Let's try to estimate the optical depth of a disk region that might form an Earth-like planet. Suppose the grains that eventually form Earth are spread out in a ring from 0.5 au–1.5 au, and that the grains are 1 μm in size. The total cross-sectional area of all of the grains is given by

$$A_{\text{grains}} = N_{\text{grains}} A_{\text{grain}}$$
$$= N_{\text{grains}} \pi R_{\text{grain}}^2. \tag{9.37}$$

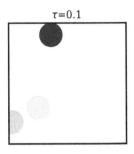 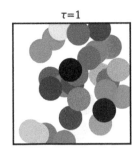 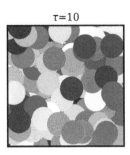

Figure 9.4. Illustration of regions with different optical depths, where we assume here that τ = (combined cross-sectional area of circles)/(area of box). Since we allow the circles (representing dust grains) to have random positions, when $\tau = 1$, the circles do not perfectly fill the space, and some light may still get through the box. Once $\tau = 10$ is reached, very little light is likely to make it through the box.

We know that it takes about 10^{38} μm-sized grains to make an Earth-like planet, so the total area is $A_{grains} \approx 10^{38} \times \pi \times (10^{-6}\,\text{m})^2 \approx 10^{26}\,\text{m}^2$. The total area of the column is given by $A_{column} = \pi(1.5\,\text{au})^2 - \pi(0.5\,\text{au})^2 \approx 10^{23}\,\text{m}^2$. Thus, the optical depth, given by the ratio A_{grains}/A_{column} is approximately 1000. The disk is *very* opaque, and virtually no light would make it through the disk.

Note that in this discussion we made the assumption that the wavelength of light is smaller than the grain size. Longer wavelengths of light are not effectively blocked by small grains; the grains end up blocking much less light than their cross-sectional areas would otherwise suggest, and we need to carefully consider how light interacts with particles to correctly calculate τ. This is also why, for example, radio waves have no trouble penetrating through the walls of your home. Because small particles don't effectively block long wavelengths of light, it turns out that at millimeter and longer wavelengths, some regions in protoplanetary disks can have $\tau < 1$.

9.2.4 An Example Disk Image

Protoplanetary disks can be imaged in nearby star-forming regions, and the images confirm that they are, indeed, geometrically thin but flared, and quite dusty. Figure 9.5 shows a disk known informally as the "flying saucer". The dark lane in this edge-on disk shows how opaque the dust can be, and demonstrates the thin, yet flared, shape expected for a protoplanetary disk. The brighter top and bottom regions are where starlight is reflecting off of the disk surface.

9.3 Solid Planet Growth

9.3.1 Grains to Planets: An Overview

The process of planet-building requires taking grains of size 1 μm and building solid bodies of size 10^6 m (Earth-sized) or larger—an increase in size of 12 or more orders of magnitude. In addition, if building a gas giant, the planet formation process must somehow aggregate gas molecules to form a massive gas atmosphere. But small grains do not act the same as large grains, and gas molecules do not act the same as solids. The process of planet formation is inherently complex, and the physical

Figure 9.5. A Hubble Space Telescope image of the protoplanetary disk known informally as the "flying saucer" (located around the much less creatively-named star 2MASS J16281370-2431391). The dark lane is the protoplanetary disk, which we can see is dusty, geometrically thin, and flared. Some starlight can escape vertically, however, and illuminates the top and bottom of the disk. Image source: ESO/NASA/ESA. CC BY 4.0.

phenomena that govern planet formation processes are numerous. Therefore, the planet formation process is often broken up into different stages—each of which is governed by a manageable number of physical principles—and the stages are pieced together to understand the full process. In this section and the next, we'll describe some of these stages, beginning with the coagulation of small grains into the building blocks of planets known as *planetesimals*, and ending with the gravitational attraction of large gas atmospheres. As in the rest of this text, we'll focus on simple physical principles and necessarily leave out many details of cutting-edge work in this field.

9.3.2 Orderly Growth and the Timescale Problem

The planet formation process begins with the coagulation of small grains into larger planetary building blocks. As shown in Figure 9.6, we can picture this process as some small sphere of radius R traveling through a swarm of grains with space density ρ_s at speed v. Each time the sphere collides with a grain, the grain sticks to the sphere, and the sphere grows—a process known as *accretion*. If the grain is large, and the density and speed high, the sphere grows quickly; if the grain is small, or the swarm density or velocity small, the sphere grows slowly. More precisely, the growth rate (rate of change of mass, m, versus time) of the sphere is given by:

$$\frac{dm}{dt} = \pi R^2 \rho_s v. \tag{9.38}$$

This simple process is called *orderly growth*, and this rate is known as the orderly growth rate. Note that as the grain grows in mass, its size, R, will increase as well,

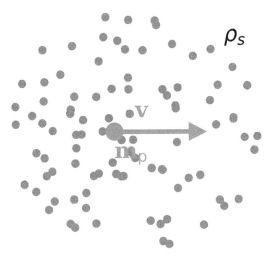

Figure 9.6. In the orderly growth scenario, a baby planet travels through a swarm of grains, growing each time it collides with a grain.

further increasing its growth rate. Thus, the growth rate is not constant, but increases with time.

We can derive a *characteristic timescale* for the orderly growth process by defining:

$$\tau_{\text{orderly}} = \frac{M_p}{dm/dt} \qquad (9.39)$$

where M_p is the final mass of the planet. Let's consider a ridiculously optimistic situation in which a net with radius R_p (the final radius of the planet) collects grains evenly distributed in the planetary growth region. (In other words, we're unrealistically allowing the planet to start at its final size, even before it has any mass.) For simplicity, let's define the planetary growth region as a torus-(i.e., donut)-like shape with width r_p (the planet's distance from the star), height equal to the disk scale height $r_p/30$, and circumference $2\pi r_p$, which has a volume of $2\pi r_p \times \pi r_p^2/30 = \frac{1}{15}\pi^2 r_p^3$. For example, for Earth, we'd be assuming a planet growth region centered at 1 au extending from 0.5–1.5 au, with a height of 1/30 au. If we spread out the mass of the planet over this volume, we'll find that

$$\rho_s = \frac{15M_p}{\pi^2 r_p^3}. \qquad (9.40)$$

Furthermore, let's assume that the velocity is given by the local orbital velocity (also known as the Keplerian velocity, v_{Kep}). With these assumptions,

$$\frac{dm}{dt} = \pi R_p^2 \frac{15M_p}{\pi^2 r_p^3} v_{\text{Kep}} \qquad (9.41)$$

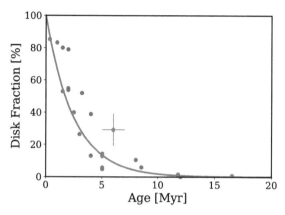

Figure 9.7. Plot of percentage of stars hosting disks versus star cluster age. (Figure and data adapted from Mamajek 2009 and references therein.) One representative error bar shows the typical uncertainty on these parameters—10%–30% on disk fraction, and ~1 Myr on disk lifetime. The solid curve shows an exponential decay with a decay constant of 2.5 Myr.

$$= R_p^2 \frac{15 M_p}{\pi r_p^3} v_{\text{Kep}} \tag{9.42}$$

and

$$\tau_{\text{orderly}} = \frac{M_p}{dm/dt}$$
$$= \frac{\pi r_p^3}{15 R_p^2 v_{\text{Kep}}}. \tag{9.43}$$

If we put in numbers relevant for the Earth ($r_p = 1$ au, $R_p \approx 6400$ km, and $v_{\text{Kep}} \approx 30$ km s^{-1}), we find that $\tau_{\text{orderly}} \approx 2 \times 10^7$ yr (20 Myr). Notice that τ_{orderly} scales steeply with distance from the star, r_p, so if, for example, we moved the Earth out to 5 au, the timescale would increase by a factor of $5^3 = 125$, and at 20 au it would increase by a factor of 8000!

Is this a reasonable growth timescale? How might we even decide such a question? Actual observations of protoplanetary disks provide significant insight. We can find out how long protoplanetary disks last by measuring the ages of nearby clusters of young stars, and asking what percentage of stars in each cluster have disks. Such a plot is shown in Figure 9.7. Notice that at the youngest ages, greater than 80% of stars host protoplanetary disks, while by ages of 10 Myr, less than 5% of stars still host a disk. If a planet forms *from* the disk dust, then the orderly growth timescales we derived above cannot be reconciled with observed disk lifetimes.

9.3.3 Gravitational Focusing and the Safronov Parameter

One important consideration we ignored in our orderly growth derivation was the effect of gravity. Given that orderly growth doesn't seem to produce planets quickly enough, let's consider whether gravity might help speed up planet growth. In the orderly growth derivation, we assumed that the growth timescale was set by the geometric cross-section of the planet. In other words, the planet had to exactly cross the path of a grain in order for the grain to accrete onto the planet. If we take gravity into account, however, there will be grains that don't exactly cross the planet's path but perhaps come close enough to get pulled toward the planet by its gravity, and eventually accrete. In this section, we'll use the principles of conservation of energy and conservation of angular momentum to derive an expression for exactly how close a grain has to get to the planet in order to accrete.

Figure 9.8 shows the setup for our derivation. A grain of mass m approaches the growing planet with a relative speed of v_∞. If there were no force of gravity, the grain would pass by the planet with a closest approach distance (measured between the centers of the two bodies) of b. This distance is known as the *impact parameter*. We'll assume that the growing planet has a mass M and radius R.

In this process, angular momentum is conserved. If there are no non-conservative forces (like drag forces, or friction) acting on the grain, then energy (kinetic plus potential) is conserved as well. Let's call the angular momentum and energy L_∞ and E_∞ when the grain is very far away, and L_{imp} and E_{imp} when the grain has just impacted the surface of the planet. Figure 9.9 shows the geometry when the grain is far from the planet. Since $\vec{L} = \vec{r} \times \vec{p}$, the magnitude of L_{inf} is given by $mrv_\infty \sin\theta = mrv_\infty(b/r) = mv_\infty b$. At impact, $L_{imp} = mRv_{imp}$. Therefore,

$$L_\infty = L_{imp} \tag{9.44}$$

$$mbv_\infty = mRv_{imp} \tag{9.45}$$

$$v_{imp} = \frac{bv_\infty}{R}. \tag{9.46}$$

Conservation of energy implies:

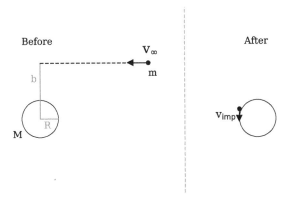

Figure 9.8. Left: A grain of mass m with speed v_∞ approaches a growing planetesimal of mass M and radius R with impact parameter b. Right: For a grazing impact, the grain just barely hits the edge of the planet.

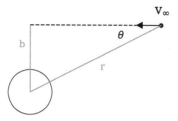

Figure 9.9. Geometry when the grain is far away from the planet.

$$E_\infty = E_{imp} \tag{9.47}$$

$$\frac{1}{2}mv_\infty^2 - \frac{GMm}{\infty} = \frac{1}{2}mv_{imp}^2 - \frac{GMm}{R} \tag{9.48}$$

$$v_\infty^2 = v_{imp}^2 - \frac{2GM}{R}. \tag{9.49}$$

We can substitute in for v_{imp} using Equation (9.46) to find:

$$v_\infty^2 = \left(\frac{bv_\infty}{R}\right)^2 - \frac{2GM}{R} \tag{9.50}$$

$$v_\infty^2\left(1 - \frac{b^2}{R^2}\right) = -\frac{2GM}{R} \tag{9.51}$$

$$1 - \frac{b^2}{R^2} = -\frac{2GM}{Rv_\infty^2} \tag{9.52}$$

$$\frac{b^2}{R^2} = 1 + \frac{2GM}{Rv_\infty^2} \tag{9.53}$$

$$b^2 = R^2\left(1 + \frac{2GM}{Rv_\infty^2}\right). \tag{9.54}$$

Since the escape speed from the growing planet is given by $v_{esc} = \sqrt{\frac{2GM}{R}}$ (recall Equation (8.9)), we can rewrite this expression in the form:

$$b^2 = R^2\left(1 + \frac{v_{esc}^2}{v_\infty^2}\right). \tag{9.55}$$

Remember that b represents the gravitational "reach" of the growing planet, so it's as if the planet has an effective gravitational size (b) that is bigger than its physical size (R). This effect is known as *gravitational focusing*. Let's look at what this expression tells us. If the planet is big (a large value of R), it'll collide with more objects—that, we already knew. However, there's now an additional term $\left(1 + \frac{v_{esc}^2}{v_\infty^2}\right)$

(with $\frac{v_{esc}^2}{2v_\infty^2}$ named the *Safronov parameter* after Viktor Safronov, who performed some of the pioneering work on this problem; Safronov 1972.) If v_∞ is very large, the Safronov parameter approaches zero, and the planet's gravitational radius b is not much different from its actual radius R. This makes intuitive sense—the grain goes whizzing by the planet, and the planet doesn't have enough gravitational pull to significantly change its course during their short interaction time. However, if v_∞ is very small, the planet's gravitational radius b grows very large, as the planet has a lot of time in which to interact with the particle. In fact, as v_∞ approaches 0, b approaches infinity!

If we revisit the orderly growth rate, dm/dt (Equation (9.38)), we can substitute in this new gravitational radius to find:

$$\frac{dm}{dt} = \pi R^2\left(1 + \frac{v_{esc}^2}{v_\infty^2}\right)\rho_s v. \tag{9.56}$$

Therefore, the growth rate of the planet is increased by $\left(1 + \frac{v_{esc}^2}{v_\infty^2}\right)$, while the characteristic growth timescale would be reduced by that same amount. If the speeds of grains (v_∞) are slow enough, therefore, the planets will grow much more quickly, helping to solve the orderly growth timescale problem.

9.3.4 Stages of Solid Planet Growth

Based on our derivation of gravitational focusing, it's clear that in order to know how quickly planets grow, we have to know the speed of the grains it's trying to accrete. As it turns out, there is no easy way to determine this speed, as it depends sensitively on many environmental parameters, including: the sizes of particles, whether there is gas around, whether there are pressure gradients in the gas, etc. In addition, each of these parameters can change as the planets grow. Therefore, the process of planet formation is largely understood with the help of computer simulations that attempt to understand the complex dynamics of the planet and surrounding grains.

In principle, computer simulations are relatively straightforward. The simulations consist of either particles or points in space (grid cells) that experience forces. Forces are computed, and particles are moved, or material moves between grid cells a little bit, according to the calculated forces. Then, the forces are recomputed, and particles and material moved again, and so on, and so on. As we calculated earlier, however, it requires tremendous numbers of grains to make planets (for example, 10^{38} μm-sized grains to make an Earth-mass planet), so creating a complete start-to-finish model of planet information would require an impractical amount of computer time to run. Instead, simulations are designed to probe particular aspects or stages of planet formation, particular regions of protoplanetary disks, etc.

In general, computer simulations have shown that planet formation proceeds at different rates at different stages, and in different locations in the disk. Depending on the local conditions and the stage of planet formation, growing planets and other materials in the disk have different speeds. Due to these different speeds, the growing planets experience different motions, different degrees of gravitational focusing, and different growth rates.

Simulations suggest that in the earliest stages of planet growth, small grains are attracted together by small pressure bumps, and gravity clumps them together into slightly larger grains. An example simulation of this process, known as the *streaming instability* (Johansen et al. 2007) is shown in Figure 9.10. Once grains reach cm–m (so-called "pebble") sizes, simulations suggest that the drag provided by the protoplanetary disk's gas slows the pebbles down, increasing the gravitational influence of any growing planets, and greatly increasing their rate of growth in a process that has come to be called *pebble accretion* (Ormel & Klahr 2010; Lambrechts & Johansen 2012). When the growing planets reach a larger size, they slow down each others' motions, increasing their relative gravitational influences. This, in turn, causes the largest planets to grow ever larger—a process known as *oligarchic growth* (Kokubo & Ida 2002). The process of oligarchic growth is shown visually in the computer simulation results of Figure 9.11. When the planets grow approximately as large as Mars, their gravitational influence on each other tends to *increase* their relative velocities, and decrease their gravitational influences. The planets then enter a slower stage of growth known as the *giant impact* stage. It's believed that in this stage of planet formation, an approximately Mars-sized object impacted the Earth, causing the formation of Earth's Moon (e.g., Canup 2004, and references therein).

Recall in Chapter 1 how the solar system's planets appear to have an even spacing, noted in Bode's law (see Equation (1.1) and Figure 1.5). As planets grow and "clear out" the debris from their vicinity, they naturally produce spaces between each other. If two planets get too close to each other, they are likely to collide and join together. The planet formation process therefore naturally produces a kind of even spacing between planets (although simulations do not seem to specifically predict the spacing given by Bode's law).

9.4 Gas Giants

So far, we've only discussed the growth of solid bodies. In this section, we'll discuss the two leading theories for the formation of gas giants—*gravitational instability* and *core accretion*.

9.4.1 Gravitational Instability and the Toomre Stability Criterion

According to the theory of *gravitational instability*, gas giants might form by directly collapsing from the disk gas—see Figure 9.12. In order for this to be the case, the gravitational force that a parcel is experiencing must "win" in a battle with any other gas motions, which would tend to pull the planetary material apart. As in the case of the Jeans mass calculation (Section 9.1.1), one such gas motion is random motion

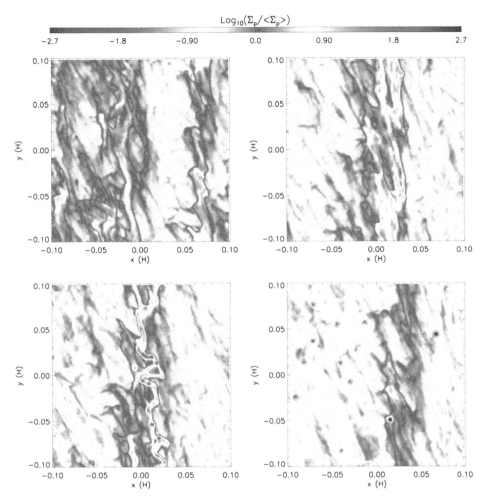

Figure 9.10. Four snapshots of a simulation of the earliest stages of planet growth, demonstrating the so-called *streaming instability*. Reproduced from Simon et al. (2016). © 2016. The American Astronomical Society. All rights reserved. The color scale represents the density of particles in the disk. [†]The x and y axes represent distances in the radial (toward and away from the star) and azimuthal (around the star) directions, respectively, and are expressed in units of the disk scale height, H. Notice how the particles collect together into higher density "streams", and then these streams allow for the formation of small clumps. [†]Strictly speaking, it's the "logarithm of the vertically integrated particle surface density normalized to the average particle surface density" (Simon et al. 2016).

due to the gas pressure. However, in a protoplanetary disk, two ends of a collapsing region would be traveling around the star at different rates according to Kepler's third law. This difference in speed, known as *Keplerian shear*, would also tend to pull the parcel apart. These three competing effects: gravity, gas pressure, and Keplerian shear, are combined into a dimensionless parameter known as the *Toomre stability criterion*, after Alar Toomre, who derived the expression (Toomre 1964). If we think about each of these effects, we might expect the expression to depend on how much

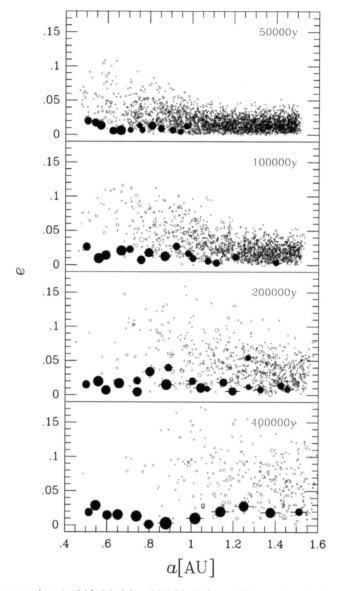

Figure 9.11. Four snapshots (at 0.05, 0.1, 0.2 and 0.4 Myr) of eccentricity and semimajor axis for growing protoplanets in a computer simulation. Reproduced from Kokubo & Ida (2002). © 2002. The American Astronomical Society. All rights reserved. The simulation begins with 10,000 equal-mass bodies of mass 1.5×10^{24} g (about 2% the mass of Earth's Moon) located between 0.5 and 1.5 au. Circles are filled when they reach 1.5×10^{26} g. The circles' sizes are proportional to the planet radii, and horizontal lines mark 10× the size of their Hill sphere (recall Equation (4.25) and Section 4.1.2). Notice how the largest bodies start to separate in mass from the rest of the objects, forming a set of "oligarchs" spanning the full range of semimajor axes.

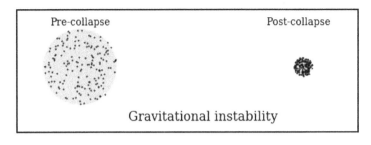

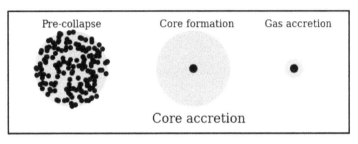

Figure 9.12. Cartoon snapshots of phases in the two proposed mechanisms for giant planet formation. In the gravitational instability scenario (top), both gas (blue) and dust (black) directly collapse from the proto-planetary disk material to form a giant planet. In the core accretion scenario (bottom), dust first collects together via collisions, aided by dynamics and gravitational focusing, to form a solid core. Then, the solid core accretes a gas atmosphere.

mass is available (more mass means more gravity), the Keplerian rotation rate (higher speeds of rotation mean greater shear) and the temperature of the gas (higher temperature means more random motions/higher pressure). The Toomre stability criterion (also known informally as the "Toomre Q") is given by the expression:

$$Q = \frac{c_s \Omega}{\pi G \Sigma} \tag{9.57}$$

where c_s is the sound speed, Ω the local Keplerian angular speed, G the gravitational constant, and Σ the local protoplanetary disk surface density (mass per unit area). We can see that Q is smallest when the mass surface density is high, while the sound speed (and thus temperature) and Keplerian rotation rate are low. Thus, smaller values of Q imply that conditions are amenable to gravitational collapse, while larger values of Q imply that the disk is stable against collapse. The boundary between the two cases occurs for $Q \approx 1$.

To get a sense of absolute values of Q, let's try computing it at 1 au in a solar nebula-like disk. Suppose the local temperature is about that of the Earth, 300 K, and the local surface density is calculated by spreading 100 Earth masses of material over a ring from 0.5–1.5 au. We're using 100 Earth masses, because we know that gas was lost during the formation of Earth, and that the original gas-to-dust ratio in the solar system was ∼100.

Then,

$$c_s = \sqrt{\frac{kT}{\mu}} \tag{9.58}$$

$$= \sqrt{\frac{(1.4 \times 10^{-23} \text{ m}^2 \text{ kg s}^{-2} \text{ K}^{-1})(300 \text{ K})}{2 \times 1.66 \times 10^{-27} \text{ kg}}} \tag{9.59}$$

$$\approx 1100 \text{ m s}^{-1} \tag{9.60}$$

$$\Sigma = \frac{100 M_\oplus}{\pi(1.5 \text{ au})^2 - \pi(0.5 \text{ au})^2} \tag{9.61}$$

$$= \frac{100 \times 6 \times 10^{24} \text{ kg}}{\pi(1.5 \times 1.5 \times 10^{11} \text{ m})^2 - \pi(0.5 \times 1.5 \times 10^{11} \text{ m})^2} \tag{9.62}$$

$$\approx 4200 \text{ kg m}^{-2} \tag{9.63}$$

and

$$\Omega = \frac{2\pi}{365 \text{ days}} \tag{9.64}$$

$$\approx 2 \times 10^{-7} \text{ s}^{-1}. \tag{9.65}$$

Therefore,

$$Q \approx \frac{(1100 \text{ m s}^{-1})(2 \times 10^{-7} \text{ s}^{-1})}{\pi(6.67 \times 10^{-11} \text{ m}^3 \text{ kg}^{-1} \text{ s}^{-2})(4200 \text{ kg m}^2)} \tag{9.66}$$

$$\approx 250. \tag{9.67}$$

Therefore, at 1 au in our hypothetical solar nebula, $Q \gg 1$ and the disk is stable against collapse (i.e., not able to form planets directly via gravitational collapse).

To calculate Q in other parts of the disk, we need to understand how Q varies with distance from the star, r. It's straightforward to calculate how Ω depends on distance from the star (since $\Omega = \sqrt{\frac{GM_*}{r^3}} \propto r^{-1.5}$) but the dependence of Σ and c_s on distance depend sensitively on the actual properties of the individual protoplanetary disk. For the minimum mass solar nebula, $\Sigma \propto r^{-1.5}$ (which you'll derive yourself as a homework problem). Disk temperatures depend on the disk structure and optical depth; we'll assume here a dependence of $T \propto r^{-0.5}$ (Andrews et al. 2009), which is the same as that for equilibrium temperature. In that case, since $c_s \propto \sqrt{T}$, $c_s \propto r^{-0.25}$.

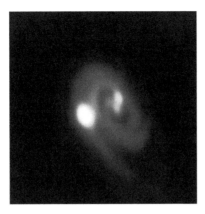

Figure 9.13. An image of a young triple stellar system, L1448 IRS3B, taken with the Atacama Large Millimeter Array (Tobin et al. 2016). The spiral structure shows a disk unstable to collapse, and the companion stars may have formed directly via gravitational instability. Image source: Bill Saxton, ALMA (ESO/NAOJ/NRAO), NRAO/AUI/NSF.

With these assumptions, $Q \propto \frac{r^{-0.25}r^{-1.5}}{r^{-1.5}} \propto r^{-0.25}$. In other words, Q decreases with distance from the star, albeit relatively slowly. For our solar nebula-like disk, even if the disk extended to 1000au, the Q value at the outer edge of the disk would still be $250 \times (1000)^{-0.25} \approx 44$, so well above the threshold for gravitational collapse.

These calculations show us that gravitational collapse is unlikely for the conditions we believed formed the planets of the solar system. However, if there are protoplanetary disks that are large, massive, and cold, Q might approach 1 at large distances from the star. Therefore, gravitational collapse may be a viable formation mechanism for some of the directly imaged planets, which are found very far from their parent stars. Figure 9.13 shows a possible example of gravitational instability in action, except that this system may have formed stars via the instabilities, rather than planets.

9.4.2 Core Accretion

An alternative proposed method for giant planet formation is known as *core accretion*. In this process, a solid core forms first via the solid growth process described in Section 9.3, and then the gravitational force of the solid body attracts a gaseous atmosphere—see Figure 9.12. Models of the process of growing the planet's atmosphere suggest that it occurs in two stages: *hydrostatic growth*, and *runaway accretion*. In the hydrostatic growth stage, hydrostatic equilibrium (Equation (8.28)) is maintained in the planet's atmosphere, and growth occurs slowly. However, once the mass of the atmosphere approaches the mass of the core, hydrostatic equilibrium can no longer be maintained, and the atmosphere collapses, entering a phase of much faster growth. An example of a model showing these two stages of planet growth is shown in Figure 9.14.

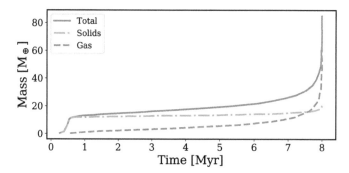

Figure 9.14. One model simulation for the growth of Jupiter via core accretion (showing mass versus time), from Pollack et al. (1996). In the first stage ($\lesssim 0.5$ Myr), the *solids* grow rapidly to form the planet's core. In the second stage (≈ 0.5–7 Myr), the core slowly begins to attract its gas atmosphere via hydrostatic growth. In the final stage ($\gtrsim 7$ Myr), the planet rapidly attracts tens of Earth masses of gas in a runaway accretion process.

The core accretion scenario naturally produces gas giant planets with solid cores surrounded by gaseous atmospheres—planets similar to Saturn, Uranus, and Neptune.[2] In contrast, gravitational instability collapses all materials at the same time, and does not naturally form a separate core and atmosphere. In addition, we estimated in the previous section that the Toomre stability criterion, Q, was likely too high in the solar nebula to form planets via direct collapse. Therefore, we believe that core accretion is the dominant mechanism of planet formation for the gas giants of the solar system.

9.4.3 Planet Types and the Snow Line

The core accretion process provides some natural explanations for the existence of the three broad classes of planets in the solar system, namely terrestrial planets (Mercury, Venus, Earth, and Mars), gas giants (Jupiter and Saturn), and ice giants (Uranus and Neptune). Ice giants Uranus and Neptune have significantly larger atmospheres than the terrestrial planets, but significantly smaller atmospheres than Jupiter and Saturn. In general, the rate of planet growth is related to the planet's Keplerian speed, since it is the planet's motion through the disk that allows it to "sweep up" the dust and gas required to feed its growth (recall Equation (9.38)). Since Keplerian orbital speeds decrease with distance from the Sun, the growth rate for Uranus and Neptune is expected to be slower than that of Jupiter and Saturn. Therefore, it may be the case that Uranus and Neptune simply grew too slowly to reach the runaway accretion stage before the gas disk dissipated (e.g., Pollack et al. 1996).

As can be seen in Figure 9.14, the process of attracting a gas atmosphere proceeds quite slowly when the planet mass is small. Therefore, it may be that the terrestrial planets simply never got massive enough to attract a significant amount of gas. However, if growth rate decreases as you get *farther* from the Sun, one may then

---

[2] Jupiter's large mass and size make it harder to determine its interior structure. However, recent results from the Juno mission (e.g., Debras & Chabrier 2019) suggest it, too, has a core, although one that is surprisingly spread out. The reasons for this are still being explored (e.g., Liu et al. 2019).

wonder—why did the terrestrial planets grow so slowly compared to the gas giants? One possibility is that the phase change of water from vapor to ice that occurs far from the Sun increases the amount of solid mass available for planet formation, thereby reducing the time needed to reach the runaway gas accretion stage. The location where this occurs is known as the disk *snow line*. In the solar system, the snow line is believed to have been near 2.5 au, based on temperature estimates (Hayashi 1981) and the locations of ice-rich versus ice-poor asteroids (Bus & Binzel 2002). This places the solar system's snow line in-between Mars and Jupiter, which makes this a compelling theory to explain the differences between terrestrial and gas giant planets.

Although the theory described in this section makes reasonable predictions for the solar system, it is not yet known whether the correspondence between the snow line location and planet type holds outside of the solar system. As we know from Chapter 5, some exoplanetary systems contain Hot Jupiters, very close to their parent stars; others contain super Earths—a type of planet not found in the solar system at all. In addition, as we will discuss in the next sections, planets can migrate far from their initial sites of formation.

9.5 Disk Evolution and Planet Migration

9.5.1 Viscosity and Disk Spreading

If gas giants form beyond the star's snow line, then why do hot Jupiters (recall Section 5.7.3) exist very close to their parent stars? Even before the discovery of hot Jupiters, theoreticians realized that a planet in a gas disk would not remain stationary, but would migrate (in a direction depending on the detailed structure of the disk, but most likely toward the star; Goldreich & Tremaine 1980; Lin & Papaloizou 1986). In addition, observations showed that disks evolved over a few Myr timescale, as discussed in Section 9.3.2 and shown in Figure 9.7. And, observations of young stars with protoplanetary disks show that hot gas is constantly falling onto the star (e.g., Muzerolle et al. 2003). Therefore, it's believed that disks are not static, but are in constant motion, spreading both toward and away from the star, falling onto the star, and thinning out over time.

The rate of spreading can be quantified in terms of a disk property known as *viscosity*. In everyday experience, viscosity is what we might think of as a kind of stickiness or resistance to flow; for example, it quantifies the difference between honey (higher viscosity) and water (lower viscosity). More precisely, this property can be quantified with ν, a fluid's kinematic viscosity, which has SI units $m^2\,s^{-1}$. For a protoplanetary disk, if viscosity is *low*, the gas molecules easily flow past each other (like molecules of water), and no evolution occurs; if viscosity is high, the gas molecules tug on each other (like molecules of honey) in some way, causing energy transfer between molecules, and the eventual spreading of the disk. Therefore, for disks, we can think of viscosity as a measure of spreadability, and of the kinematic viscosity as the area over which the disk spreads per unit time; a highly viscous disk spreads rapidly, while a low-viscosity disk spreads hardly at all. Note that this makes *disk* viscosity somewhat opposite to our everyday understanding of viscosity, since a drop of viscous honey on a table will spread much less rapidly than water, while a drop of high viscosity *disk* will spread *more* rapidly than a drop of low-viscosity disk.

For honey, the higher viscosity is provided by intermolecular electromagnetic forces. For protoplanetary disks, the underlying physics is not yet clear, but may be due to the presence of magnetic fields. Nevertheless, we can obtain a rough estimate of a disk's kinematic viscosity by considering its size and evolution rate, even if we have no idea what physics is causing it. If disks are about $R \approx 100$ au in size, and take $\tau \approx 3$ Myr to evolve (see Figure 9.7), then, remembering that viscosity is the area of spread over time (Ruden 1999),

$$\nu \approx \frac{R^2}{\tau} \tag{9.68}$$

$$\approx \frac{(100 \text{ au})^2}{3 \text{ Myr}} \tag{9.69}$$

$$\approx 2 \times 10^{12} \text{ m}^2 \text{ s}^{-1}. \tag{9.70}$$

For context, this value is much higher than both water and honey (which both have $\nu \approx 10^{-5}$ m^2 s^{-1}) but is much lower than the viscosity of the rocks in Earth's mantle (for which $\nu \approx 10^{22}$ m^2 s^{-1}).

We can also invert Equation (9.68) to find a viscous timescale:

$$\tau \approx \frac{R^2}{\nu}. \tag{9.71}$$

Since this scales as R^2, we can use proportional reasoning to estimate the evolution timescale in the planet-forming regions of the disk. At 5 au, for example, the timescale would be

$$\tau_{5 \text{ au}} \approx (3 \text{ Myr}) \times \left(\frac{5 \text{ au}}{100 \text{ au}} \right)^2 \tag{9.72}$$

$$\approx 7500 \text{ yr}, \tag{9.73}$$

and at 1 au, it would be only 300 yr. But are planets affected by this disk spreading process?

9.5.2 Migration of Giant Planets (Type II Migration)

As it turns out, planets are, indeed, affected by the evolution of the disk. In particular, gas giant planets open gaps surrounding their orbits, and then effectively get trapped in those gaps, and migrate toward the star on the viscous timescale. Why do giant planets open gaps? Due to the sometimes non-intuitive physics of planetary orbits, the gas giants' gravitational forces push the gas away from them, in a process known as shepherding.

The process of shepherding can be understood by considering the diagrams shown in Figures 9.15 and 9.16. Figure 9.15 shows a planet moving through a

Stationary Frame

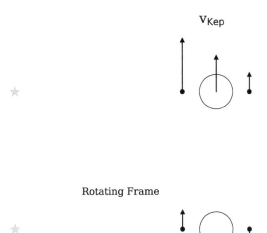

Rotating Frame

Figure 9.15. Top: An exaggerated comparison of orbital speeds for a planet and two adjacent particles orbiting a star in a counterclockwise direction. Kepler's third law tells us that orbital speed decreases with distance from the star. Bottom: In a rotating frame of reference orbiting around the star at the same rate as the planet, interior particles appear to pass by moving upwards, while exterior particles pass by moving downwards. In both panels, we are viewing the protoplanetary disk from above.

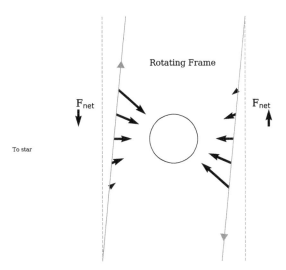

Figure 9.16. An exaggerated view of the interaction between an orbiting planet (or moon) and nearby particles, as viewed in the rotating frame. As a particle passes by the planet, the gravitational forces cause it to approach the planet. Therefore, the gravitational interactions are stronger during the second half of the interaction. Interior particles feel a net force downwards, while exterior particles feel a net force upwards. Since both particles are *actually* moving upwards (in the inertial frame), the exterior particle experiences positive work and gains energy, while the interior particle experiences negative work and loses energy.

protoplanetary disk, along with the paths of two adjacent particles. The arrows show an exaggerated comparison of their speeds, which are faster closer to the star and slower farther from the star, according to Kepler's third law. If we move our perspective to a frame of reference rotating around the star at the same speed as the planet, we get the situation viewed in Figure 9.16. The outer particle appears to move downwards relative to the planet, while the inner particle appears to move upwards. Let's first consider the gravitational effect of the planet on the outer particle. As the particle moves downwards, it first feels a net force partially downwards, and then a net force partially upwards (as viewed in the rotating frame). However, since the gravitational force also pulls the particle closer to the planet, the second half of the interaction is slightly stronger, and the particle feels slightly more force upwards, making the overall interaction a slight upwards push. Although we considered this interaction in the rotating frame, the particle is actually moving upwards, so this upwards push does positive work on the particle, and gives it a small energy boost. Since larger orbits have greater energy (since, as we showed in Equation (3.57), $E = -\frac{Gm_\odot m_p}{2a}$), the energy boost moves the particle to a larger orbit, *away* from the planet. Therefore, it's as if the gravity of the planet is pushing the particle away! A similar effect occurs for the particle in the interior orbit. As shown in Figure 9.16, the particle moves upwards past the planet, and feels a slight net force downwards, opposite to its direction of motion. Thus, the gravitational force of the planet does negative work on the particle, the particle loses energy, and it moves to a smaller orbit, again *away* from the planet. Once again, it appears as if the gravitational force of the planet is acting to repel the particle.

The process of shepherding can be seen in our own solar system—it is the process by which moons carve gaps in Saturn's rings. An example is shown in Figure 9.17. In

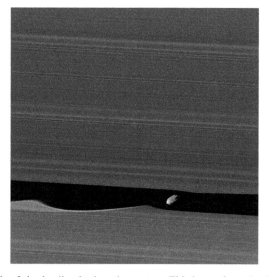

Figure 9.17. An example of shepherding in the solar system. This image from the Cassini spacecraft shows Saturn's moon Daphnis carving a gap in the rings of Saturn. Image source: NASA/JPL-Caltech/Space Science Institute.

a protoplanetary disk, the planet's gravitational force works to push gas away from the planet, while the random motions of the gas push back on the planet. Therefore, only large, Jupiter-like planets, have enough gravitational force to keep the gas at bay, and open a gap. Once the planet opens the gap, these continual pushes keep the planet trapped in the gap even as the disk evolves, and the planet moves toward the star as the disk viscously evolves. This process of gap opening followed by migration toward the star is known as *Type II migration*,[3] and is a proposed mechanism by which hot Jupiters can form beyond the snow line and end up close to their parent stars. However, the origin of hot Jupiters is still being investigated. In addition, it is not known for sure why Jupiter apparently did NOT migrate toward the Sun, but remained near the snow line.

Recent observations of protoplanetary disks have also begun to show evidence for shepherding. Figure 9.18 shows millimeter-wave images of 12 protoplanetary disks in the Taurus star-forming region (Long et al. 2018). Many of these disks have gaps and rings that are believed to be opened via planet shepherding.

9.5.3 Signatures of Early Solar System Dynamics

While Jupiter obviously did not migrate all the way to the Sun, the solar system does have some clear (as well as more subtle) evidence for a dynamic history. Some observations that have been invoked as evidence for past migration include: the proximity of Jupiter and Saturn to the 3:2 or 2:1 resonance, the small mass of Mars, and the absence of super-Earths in the solar system. But perhaps the clearest piece of evidence for migration is the structure of the so-called Kuiper Belt.

As discussed in Chapter 4, the Kuiper Belt is a collection of objects orbiting past Neptune. Pluto is a member of this collection, but it was not known that Pluto

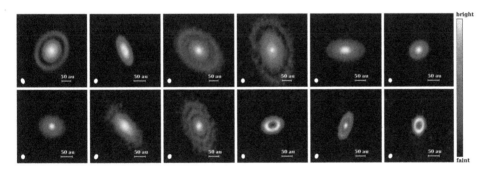

Figure 9.18. Images of protoplanetary disks in the Taurus star-forming region taken by the Atacama Large Millimeter Array, with many showing evidence of shepherding of disk material by planets. Reprinted from Long et al. (2018). © 2018. The American Astronomical Society. All rights reserved.

---

[3] Less massive planets may also migrate through the disk, driven by a different physical process known as Type I migration.
[4] Or third object, if you count Pluto's moon Charon.

wasn't unique until the discovery of the second object[4] (named Albion) in 1992 (Jewitt & Luu 1993). Since then, hundreds of Kuiper Belt objects have been discovered, and the existence of a much larger number of objects too small to detect can be inferred.

Figure 4.8 shows eccentricity versus semimajor axis for known objects in the Kuiper Belt (referred to as Kuiper Belt objects, or as trans-Neptunian objects). Note that there are a large number of objects beyond ~47 au that have large eccentricities, but few that have low eccentricities. In addition, note the large number of objects in mean motion resonances (MMRs) with Neptune. These features suggest a process in which Neptune migrated outwards, dynamically *pushing* Kuiper Belt objects outwards (e.g., Morbidelli & Nesvorný 2020, and references therein). During this process, many objects end up trapped, in a sense, by Neptune's gravity, forced to encounter Neptune over and over again according to the ratio of their periods. They end up in MMRs with Neptune. Other objects get gravitationally perturbed as they get pushed outwards, ending up with high eccentricities, inclinations, and semimajor axes.

Because we have so much detailed information about the solar system, some scientists have tried constructing scenarios that can fully explain the locations, dynamics, and sizes of solar system bodies. One such model, which has come to be called the "Nice model" (after the French town, not the adjective; Gomes et al. 2005; Morbidelli et al. 2005; Tsiganis et al. 2005), begins with the giant planets closely packed together, and proceeds to a disruptive event in which the planets migrate outwards, Uranus and Neptune switch places, and small bodies are flung inwards to cause the Late Heavy Bombardment (see Chapter 7). The details of this model continue to be explored and tweaked to explain our solar system, but one fact remains clear—the early solar system was a dynamic place.

9.6 Meteorites: Cosmochemical Probes and Planet Formation Clocks

9.6.1 Types and Origins of Meteorites

The study of meteorites (rocks that have fallen to Earth from space) provides us with an incredibly detailed, nuanced picture of the early solar system. Unlike with solar system planets, extrasolar planets, or protoplanetary disks, we can actually hold a meteorite in our hand, and therefore analyze it in our lab, making precise measurements of bulk, isotopic, and mineral compositions. The details provided by meteoritics in turn present a unique challenge—how to reconcile incredibly detailed information about the early solar system with a single formation scenario. We won't attempt to address or tackle any difficult questions here, but will provide just an overview of key ideas. We encourage an interested reader to delve into the rich literature of meteoritics for more details.

At the highest level, meteorites come in two types: primitive and differentiated. Primitive meteorites, known as chondrites, were never melted, and therefore are believed to closely represent the building blocks of larger planets. Differentiated meteorites (often called achondrites, although naming conventions vary) are the opposite, showing chemical evidence for past melting. These meteorites represent pieces of objects that differentiated, and later broke apart. They come in three subtypes: irons, stones, and stony-irons, believed to represent the core, mantle, and core–mantle boundary of their parent bodies.

Aside from these basic classifications, chemical analyses can place meteorites into distinct groups, which are believed to each map to a single parent body, or to a small number of very similar parent bodies. The chemistry of each group of meteorites can tell us the age of the parent body (discussed in Section 9.6.3), the extent to which the body was melted, and where it may have formed.

9.6.2 Chondrites and the Condensation Sequence

If chondrites are primitive, i.e., unaltered, their compositions should tell us about the composition of the solar nebula. In fact, if we measure the bulk chemistry of

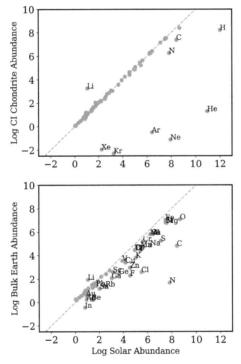

Figure 9.19. Atomic abundances in primitive chondrite meteorites (called "CI"; top) and the Earth (bottom) as compared to atomic abundances in the Sun. (Data from Grevesse et al. 2010; Lodders et al. 2009; Allègre et al. 2001.) Dashed lines mark a 1:1 correspondence, and elements with large deviations from this line are labeled. Abundances are normalized assuming the solar abundance of H is 10^{12}. For the Earth, the abundance of refractory molybdenum is taken to be the same as that in the Sun. Note how well abundances match in CI chondrites and the Sun, except for a few elements that are found in the gas phase even at low temperatures. CI chondrites are believed to have formed in the outer asteroid belt (beyond 4 au). However, for the Earth, many more elements are depleted as compared to the Sun (and the noble gases are below the axes). Lithium is unusual for being depleted in the Sun, which is a result of the fusion process. In fact, Lithium abundances are used as a means to measure stellar ages. Pontoppidan et al. (2014).

chondrites, we find something remarkable (see Figure 9.19): for all but the lightest atoms (which tend to be in the gas phase), the composition of chondrites match the composition of the Sun. Thus, chondrites are indeed nearly pristine pieces of the solar nebula.

The tendency of an atom to prefer to be in the gas phase or the solid phase is known as its *volatility*, and depends on the condensation temperature of its major molecular (or atomic) carrier. For example, He and other noble gases remain as single atoms, and have very low condensation temperatures, preferring to remain in the gas phase for most temperatures. Such atoms are termed *volatile* elements. In contrast, Fe and other metals have very high condensation temperatures, preferring to remain in the solid phase for most temperatures. These atoms are termed *refractory* elements. Atoms like N, C, and O are often found in the molecules NH_3, CH_4, and H_2O—molecules that we often call "ices". These molecules have intermediate condensation temperatures, and in the solar system, they are sometimes found in the gas phase, and sometimes in the solid phase (or at higher pressure, as in the case of the oceans on Earth, in the liquid phase).

An interesting trend emerges if we look at the elemental composition of objects in the solar system. The closer the object is to the Sun, the fewer volatile elements it has. For example, the bottom half of Figure 9.19 shows our best estimate of the elemental composition of the Earth. As compared to the chondrite abundances, we see that H has been greatly depleted, the noble gases are so depleted as to be off of the chart, and elements with moderate volatility, like C, N, and O are also less common. The chondrite shown here was believed to have formed at 4 au, while the Earth, as we know, is at 1 au. And, the chondrite has more volatile elements than the Earth. This trend continues if we consider the compositions of both the planets of the solar system, and of meteorites.

What is the origin of this trend? This compositional gradient can be reproduced with a scenario known as the *condensation sequence*. In this scenario, the early solar system is initially very hot, and well-mixed, so that it is fully vaporized, and the elemental composition is the same everywhere. As time proceeds, the solar system cools. The most refractory elements condense into molecules first, forming refractory-rich grains, and as the solar system cools further, more volatile elements begin to condense (Grossman 1972). The final temperature of any given location, however, is determined by its distance from the Sun. Mercury, close to the Sun, forms in a high-temperature environment, and only very refractory elements condense out near its location in the solar nebula. Anything remaining in the gas phase is not readily incorporated into Mercury, as Mercury is not massive enough to gravitationally hold onto a large atmosphere. Therefore, Mercury becomes a refractory-rich planet. Venus, being farther from the Sun, has more volatile elements than Mercury, the Earth has more volatile elements than Venus, and so on. In the outer solar system, O is able to condense as H_2O, and become part of the building blocks of planetary cores. As discussed in 9.4.3, the location where H_2O condenses out is known as the snow line, and the availability of water ice makes Jupiter's core grow large enough to enable runaway gas accretion. Finally, bodies in the outer solar system incorporate ices, like NH_3, CH_4, and even frozen N_2.

9.6.3 ^{26}Al Decay and the Early Solar System Timeline

Meteorites also serve as a kind of clock for solar system formation. As discussed in Section 6.3.2 radioactive decay is the spontaneous decay of parent atoms into daughter atoms, with an accompanying release of energy. A particularly important radioactive element in the early solar system was ^{26}Al. Given that the planet formation timescale is thought to be about a few Myr, and the half-life of ^{26}Al is 0.72 Myr, this radionuclide would have been decaying readily during the planet formation process. This not only provided energy to young planets as they formed (see Section 6.3.2) but allows us to use this decay process as a means to construct a timeline of early solar system events, using *radiometric dating*.

The basic technique is as follows: we assume the solar system started with an injection of ^{26}Al. Over time, the building blocks of planets form, while the amount of ^{26}Al decays. If objects form early, they'll form with a lot of ^{26}Al, while if they form later, they'll form with little (or no) ^{26}Al. But how do we measure how much ^{26}Al was originally in the object, if it will later all decay to ^{26}Mg? The secret is to look for the decay product, ^{26}Mg—extra ^{26}Mg today implies extra ^{26}Al at formation.

This is a good idea, but suffers from one serious complication. ^{26}Mg is not only produced by decay, but would exist in the object anyway. How do we quantify any "excess" ^{26}Mg that was due to decay only? To quantify how much ^{26}Mg was due to decay, versus originally in the body, we don't measure absolute amounts of ^{26}Mg, but rather amounts of ^{26}Mg relative to the stable isotope ^{24}Mg. If there appears to be an excess of ^{26}Mg compared to ^{24}Mg, we can infer that the body originally had some ^{26}Al that decayed to ^{26}Mg.

Let's do some math to see how this works. Let ^{26}Mg represent the total amount of ^{26}Mg in the body today, and let ^{26}Mg$_0$ and ^{26}Al$_0$ be the original amounts of ^{26}Mg and ^{26}Al in the bodies. These quantities are related as follows:

$$^{26}\text{Mg} = {}^{26}\text{Al}_0 + {}^{26}\text{Mg}_0. \qquad (9.74)$$

If we divide the equation by the amount of the stable isotope ^{24}Mg, then:

$$\frac{^{26}\text{Mg}}{^{24}\text{Mg}} = \frac{^{26}\text{Al}_0}{^{24}\text{Mg}} + \frac{^{26}\text{Mg}_0}{^{24}\text{Mg}}. \qquad (9.75)$$

Now, we'll use another stable isotope: ^{27}Al. We can multiply the first term on the right-hand side by $\frac{^{27}\text{Al}}{^{27}\text{Al}}$, since this is just equal to 1, and then rearrange the ratios to find:

$$\frac{^{26}\text{Mg}}{^{24}\text{Mg}} = \frac{^{26}\text{Al}_0}{^{27}\text{Al}} \cdot \frac{^{27}\text{Al}}{^{24}\text{Mg}} + \frac{^{26}\text{Mg}_0}{^{24}\text{Mg}}. \qquad (9.76)$$

If we have a meteorite sample in the lab, we can measure the ratios $\frac{^{26}\text{Mg}}{^{24}\text{Mg}}$ and $\frac{^{27}\text{Al}}{^{24}\text{Mg}}$ for each sample, but we do not know the values of $\frac{^{26}\text{Mg}_0}{^{24}\text{Mg}}$ or $\frac{^{26}\text{Al}_0}{^{27}\text{Al}}$. However, if we have many samples, each with different $\frac{^{26}Mg}{^{24}Mg}$ and $\frac{^{27}Al}{^{24}Mg}$ ratios, we can plot these values, and they will form a line with slope $\frac{^{26}Al_0}{^{27}Al}$ and intercept $\frac{^{26}Mg_0}{^{24}Mg}$. In other words, Equation (9.76) can be seen as the equation for a line, $y = mx + b$, where $y = \frac{^{26}Mg}{^{24}Mg}$, $m = \frac{^{26}Al_0}{^{27}Al}$, $x = \frac{^{27}Al}{^{24}Mg}$ and $b = \frac{^{26}Mg_0}{^{24}Mg}$. Since the slope provides the original amount of ^{26}Al in the body, the steeper the slope of the line, the older the body is.

The oldest found objects in the solar system—see Figure 9.20—are known as calcium aluminum-rich inclusions (CAIs), due to their compositions. Figure 9.21 shows how the Al and Mg ratios are plotted to find the amount of initial ^{26}Al in such

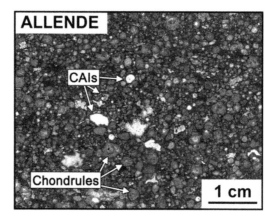

Figure 9.20. A small piece of the Allende meteorite, showing calcium aluminum-rich inclusions (CAIs)—the oldest objects formed in the solar system, and chondrules—small spherules found throughout meteorites. Reprinted from MacPherson & Boss (2011), Copyright (2011), with permission from the National Academy of Sciences.

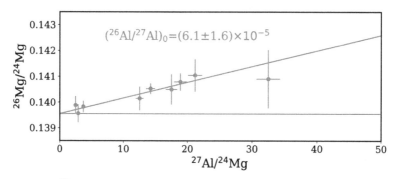

Figure 9.21. Example ^{26}Al age-dating data for a calcium aluminum-rich inclusion (CAI) called Semarkona 1805-2-1. If there were no ^{26}Al present when the rock formed, the data would lie along the horizontal line. Instead, the data have a non-zero slope equal to the rock's $\frac{^{26}Al_0}{^{27}Al}$ ratio. The amount of ^{26}Al is higher in CAIs than in all other meteoritic materials, suggesting they are the oldest solar system materials. (Data from Huss et al. 2001; Russell et al. 1996.)

a sample. The relatively high value for $\frac{^{26}Al_0}{^{27}Al}$ implies that this object formed early in the solar system.

In addition to ^{26}Al, many other radioactive isotopes are used to date early solar system materials. All such methods rely on certain assumptions, including a common start time to the solar system, uniform distributions of the isotope throughout the solar system, no loss of the parent or daughter during the meteorite's lifetime, and accurate lab measurements of the isotopic quantities and half-lives.

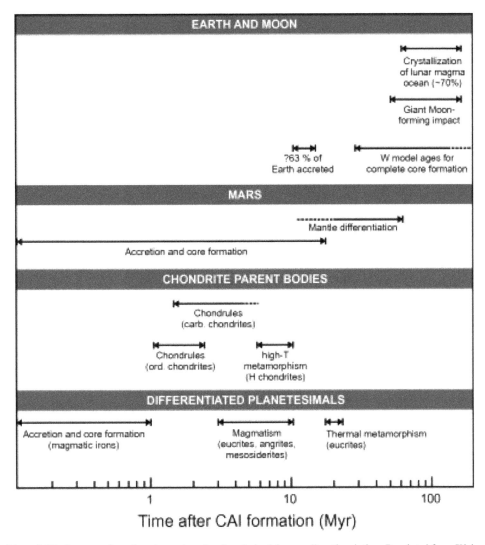

Figure 9.22. An example early solar system timeline derived from radioactive dating. Reprinted from Kleine et al. (2009), Copyright (2009), with permission from The Geochemical Society and The Meteoritical Society. Note that words in parentheses are different classes of meteorites, and that "W model ages" refers to age dating using the element Tungsten, for which ^{182}W is a decay product of ^{182}Hf.

Therefore, having multiple isotopic systems provides a way to double-check the measurements, in case any of those assumptions break down, or there are any systematic measurement errors. In addition, since each of the isotopes has a different half-life, each will be sensitive to a different timescale in solar system history.

Over time, meteoriticists have been able to use radiometric dating to construct a timeline for solar system history. An example timeline is shown in Figure 9.22. Note that the *x*-axis is given as "time after CAI formation", since CAIs are the oldest solar system materials, and all ages are given relative to those materials. In general, the timeline shows the formation of small bodies, followed by accretion into larger bodies, followed by core formation. However, one perhaps surprising result has come from these techniques as well; note that "magmatic irons" (iron meteorites), which are believed to come from the interiors of differentiated objects, are actually older than undifferentiated chondrules and chondrites. This suggests that some planet formation happened very early, and that planet formation may have proceeded at different rates in different parts of the solar system—perhaps due to the different growth processes discussed in Section 9.3.4. The formation of differentiated bodies early in the solar system's history may also have been aided by the heating provided by abundant radioactive isotopes, including ^{26}Al.

9.7 Important Terms

- Jeans mass;
- protoplanetary disk/circumstellar disk;
- solar nebula;
- gas-to-dust ratio;
- Minimum Mass Solar Nebula (MMSN);
- optical depth;
- planetesimals;
- accretion;
- orderly growth;
- characteristic timescale;
- impact parameter;
- gravitational focusing/Safronov parameter;
- streaming instability;
- pebble accretion;
- oligarchic growth;
- giant impacts;
- gravitational instability;
- Keplerian shear;
- Toomre stability criterion;
- core accretion;
- hydrostatic growth;
- runaway accretion;
- snow line;
- viscosity;
- Type II migration;

- refractory/volatile;
- condensation sequence;
- half-life;
- extinct radionuclide;
- radiometric dating.

9.8 Chapter 9 Homework Question

1. **Jeans mass:**[5]

   The Jeans mass, given in Equation (9.10), is an estimate of the minimum mass required for a gas cloud to collapse due to gravity. In this question, you'll plug some numbers into this equation to see what the implications might be.

   (a) Let's assume that a typical gas cloud has $T \approx 10$ K and $\rho \approx 10^{-20}$ kg m^{-3} (Miville-Deschênes et al. 2017), and is composed primarily of hydrogen. Using these properties, calculate the Jeans mass. (In reality, gas clouds have a range of properties, so this just gives us a ballpark idea.)

   (b) Add up the mass of the planets in the solar system and compare to the mass of the Sun. If someone asked you to calculate the mass of the solar system, do you think it would be important to include the planets in your answer? Explain.

   (c) Compare the Jeans mass to the mass of the solar system. What did you find? Explain how you might reconcile these two numbers and/or what this might tell you about planet and star formation.

2. **Angular Momentum:**[5]

   (a) Estimate the angular momentum of the solar system planets, by assuming they are point masses in circular orbits, and ignoring their spins.

   (b) Calculate the angular momentum of the Sun's spin. Assume it rotates as a solid body and has a moment of inertia factor, α, of 2/5 (a uniform sphere). (In reality, the Sun is neither a uniform sphere, nor does it undergo solid body rotation, but this calculation will at least give us a sense of the magnitude of the Sun's angular momentum.)

   (c) Based on your calculations, what contains the bulk of the solar system's angular momentum—the Sun, or the planets? Discuss your answer, and any possible implications.

   (d) Estimate the angular momentum of a spinning molecular cloud. Assume the density given in Question 1, a radius of 30 pc, and that the cloud undergoes solid body rotation with a velocity of ~5 km s^{-1} (Miville-Deschênes et al. 2017). (In reality, molecular clouds have a range of sizes, and they have both orbital and turbulent motions.)

---

[5] Adapted from coursework for Caltech course Ge/Ay 133.

(e) How do your answers to (a), (b) and (d) compare? What might this tell you about the formation of planets and stars? Make sure to consider the concept of conservation of angular momentum.

3. **The minimum mass solar nebula:**[5]

In 1981, scientist Chushiro Hayashi (Hayashi 1981) asked—given what we know about the solar system planets, can we back out what the solar system might have looked like before the planets formed? In particular, he used the positions and masses of the solar system planets to compute the surface density (mass per unit area) versus distance from the Sun for a hypothetical primordial protoplanetary disk. This has come to be called the Minimum Mass Solar Nebula (MMSN). In this question, we'll try to reproduce his work.

(a) First, the presence of gas giants, and our observations of other protoplanetary disks, tell us that the early solar system had a lot more gas than it does now. In addition, outer solar system bodies contain a lot of ice, while inner solar system bodies do not. Therefore, the inner solar system must have originally had more ice. Using measured solar abundances, and some simple assumptions, estimate the expected early solar system ratios of ice mass to rock mass, and gas mass to rock mass. Solar abundances (# relative to H) are provided below. You should also assume that all of the heavier elements (Mg, Fe, Si) are locked up into common minerals like olivine (Mg_2SiO_4, Fe_2SiO_4) and quartz (SiO_2), and that the N and C and remaining O are primarily locked up in ice molecules like H_2O, CH_4 (methane) and NH_3 (ammonia). (Hint: You may find it helpful to consider a fixed number of hydrogen atoms, say 10^6, and then ask how many molecules of olivine you'd expect to have for every 10^6 Hydrogen atoms. Don't forget to convert from number ratios to mass ratios. Finally, make sure to do proper accounting for atoms, like O, that appear in more than one molecule.)

(b) As discussed above, to determine the MMSN, we need to first account for any "missing" ice or gas in the current solar system. Using the table below, which shows the rock masses of the planets, and your gas/rock and ice/rock mass ratios, compute the "original" masses of the planets (i.e., the total original mass required to form the planet, before any ice or gas were lost). Let's also include the asteroid belt as a "planet", with position 3 au and mass 2×10^{21} kg (Pitjeva & Pitjev 2018).

(c) Using your computed original masses, compute the *total* original mass of the solar system planets, and express it in units of solar masses ($M_\odot$).

Atom	Solar Abundance by # Relative to H
H	1
He	9×10^{-2}
O	5×10^{-4}
C	3×10^{-4}
N	7×10^{-5}
Mg	4×10^{-5}
Si	3×10^{-5}
Fe	3×10^{-5}

From Grevesse et al. (2010).

Planet	Rock Mass	Original Mass	Ring Location (Inner, Center, Outer)	Surface Density [kg m^{-2}]
Mercury	3×10^{23} kg		(, 0.4 au,)	
Venus	5×10^{24} kg		(, 0.7 au,)	
Earth	6×10^{24} kg		(, 1 au,)	
Mars	6×10^{23} kg		(, 1.5 au,)	
Asteroid belt	2×10^{21} kg		(, 3 au,)	
Jupiter	1×10^{26} kg		(, 5 au,)	
Saturn	1×10^{26} kg		(, 10 au,)	
Uranus	6×10^{24} kg		(, 20 au,)	
Neptune	6×10^{24} kg		(, 30 au,)	

(d) How does the original mass of the solar system planets compare to the total mass of the planets today? Discuss your answer, including what this might say about the efficiency of planet formation.

(e) To construct the protoplanetary disk out of which the solar system formed, we can imagine the original masses of the planets spread out into a series of rings, which serve as the "feeding zones" for each planet. A reasonable assumption might be that the feeding zones extend halfway to the next adjacent planet. So, for example, Saturn's feeding zone inner edge might extend halfway to Jupiter, and its outer edge might extend halfway to Uranus. Compute the inner and outer edges of each ring and add them to the chart. Then, calculate the surface density Σ (mass divided by surface area) of each ring.

(f) Make a plot of the *log* of surface density versus the *log* of distance from the Sun (in units of au). Find and plot a line that roughly matches the data (not including the asteroid belt, which you'll find is an outlier). Does the surface density of the MMSN increase or decrease with distance from the Sun?

(g) By what factor does the region near the asteroid belt appear to be depleted? What are the implications of this result? (Hint: Think carefully about what we did and did not account for when considering mass lost during the planet formation process in part (a).)

(h) Comment on how/why the construction of this MMSN (its total mass, as well as its distribution) might be useful for planet formation studies.

(i) Although the MMSN is a very useful tool, it also has limitations. Considering the process you took to calculate the MMSN, discuss what you think may be some of its limitations, and reasons to be cautious when using it. In your answer, you should also discuss why this is called the *minimum* mass solar nebula.

References

Allègre, C., Manhès, G., & Lewin, É. 2001, E&PSL, 185, 49

Andrews, S. M., Wilner, D. J., Hughes, A. M., et al. 2009, ApJ, 700, 1502

Bus, S. J., & Binzel, R. P. 2002, Icar, 158, 146

Canup, R. M. 2004, PhT, 57, 56

Debras, F., & Chabrier, G. 2019, ApJ, 872, 100

Goldreich, P., & Tremaine, S. 1980, ApJ, 241, 425

Gomes, R., Levison, H. F., Tsiganis, K., et al. 2005, Natur, 435, 466

Grevesse, N., Asplund, M., Sauval, A. J., et al. 2010, Ap&SS, 328, 179

Grossman, L. 1972, GeoCoA, 36, 597

Hayashi, C. 1981, PThPS, 70, 35

Huss, G. R., MacPherson, G. J., Wasserburg, G. J., et al. 2001, M&PS, 36, 975

Jewitt, D., & Luu, J. 1993, Natur, 362, 730

Johansen, A., Oishi, J. S., Mac Low, M.-M., et al. 2007, Natur, 448, 1022

Kleine, T., Touboul, M., Bourdon, B., et al. 2009, GeoCoA, 73, 5150

Kokubo, E., & Ida, S. 2002, ApJ, 581, 666

Lambrechts, M., & Johansen, A. 2012, A&A, 544, A32

Lin, D. N. C., & Papaloizou, J. 1986, ApJ, 309, 846

Liu, S.-F., Hori, Y., Müller, S., et al. 2019, Natur, 572, 355

Lodders, K., Palme, H., & Gail, H.-P. 2009, Landolt–Börnstein—Group VI Astronomy and Astrophysics, Vol. 4B (New York: Springer), 712

Long, F., Pinilla, P., Herczeg, G. J., et al. 2018, ApJ, 869, 17

MacPherson, G. J., & Boss, A. 2011, PNAS, 108, 19152

Mamajek, E. E. 2009, in AIP Conf. Proc. 1158, Exoplanets and Disks: Their Formation and Diversity (Melville, NY: AIP), 3

McCaughrean, M. J., & O'dell, C. R. 1996, AJ, 111, 1977

Miville-Deschênes, M.-A., Murray, N., & Lee, E. J. 2017, ApJ, 834, 57

Morbidelli, A., Levison, H. F., Tsiganis, K., et al. 2005, Natur, 435, 462

Morbidelli, A., & Nesvorný, D. 2020, The Trans-neptunian Solar System (Amsterdam: Elsevier), 25

Muzerolle, J., Hillenbrand, L., Calvet, N., et al. 2003, ApJ, 592, 266

Ormel, C. W., & Klahr, H. H. 2010, A&A, 520, A43

Pitjeva, E. V., & Pitjev, N. P. 2018, AstL, 44, 554

Pollack, J. B., Hubickyj, O., Bodenheimer, P., et al. 1996, Icar, 124, 62

Pontoppidan, K. M., Salyk, C., Bergin, E. A., et al. 2014, Protostars and Planets VI (Tucson, AZ: Univ. of Arizona Press), 363

Ruden, S. P. 1999, in NATO ASI Series C 540, The Origin of Stars and Planetary Systems, ed. C. J. Lada, & N. D. Kylafis (Dordrecht: Kluwer), 643

Russell, S. S., Srinivasan, G., Huss, G. R., et al. 1996, Sci, 273, 757

Safronov, V. S. 1972, Evolution of the protoplanetary Cloud and Formation of the Earth and Planets (Translated from Russian) (Jerusalem: Israel Program for Scientific Translations, Keter Publishing House)

Simon, J. B., Armitage, P. J., Li, R., et al. 2016, ApJ, 822, 55

Tobin, J. J., Kratter, K. M., Persson, M. V., et al. 2016, Natur, 538, 483

Toomre, A. 1964, ApJ, 139, 1217

Tsiganis, K., Gomes, R., Morbidelli, A., et al. 2005, Natur, 435, 459

Introductory Notes on Planetary Science
The solar system, exoplanets and planet formation
Colette Salyk and Kevin Lewis

Recommended Texts for Further Reading

We hope you've enjoyed this introduction to the physics of planets. We have inevitably left many topics, and many details, out of this text. We include here a list of recommended texts, should a reader wish to continue their studies of planetary science.

Armitage, P. J. 2007, Lecture Notes on The Formation and Early Evolution of Planetary Systems, arXiv: astro-ph/0701485

Beatty, J. K., Collins Petersen, C., & Chaikin, A. (ed) 1999, The New Solar System (Cambridge: Cambridge Univ. Press)

Chamberlain, J. W., & Hunten, D. M. 1987, Theory of Planetary Atmospheres. An Introduction to their Physics and Chemistry, International Geophysics Series (Vol. 36,; Orlando, FL: Academic)

Cole, G. H. A., & Woolfson, M. M. 2002, Planetary Science: The Science of Planets Around Stars (Bristol: The Institute of Physics Publishing)

de Pater, I., & Lissauer, J. J. 2015, Planetary Sciences (Cambridge: Cambridge Univ. Press)

Hartmann, W. K. 2005, Moons and Planets (5th ed.; Belmont, CA: Thomson Brooks/Cole)

Houghton, J. 2002, The Physics of Atmospheres (Cambridge: Cambridge Univ. Press)

Ingersoll, A. P. 2013, Planetary Climates (Princeton, NJ: Princeton Univ. Press)

Johnson, J. A. 2015, How do you Find an Exoplanet? (Princeton, NJ: Princeton Univ. Press)

Melosh, H. J. 2011, Planetary Surface Processes, Cambridge Planetary Science (13) (Cambridge: Cambridge Univ. Press)

Ruden, S. P. 1999, in NATO ASI Series C 540, The Origin of Stars and Planetary Systems, ed. C. J. Lada, & N. D. Kylafis (Dordrecht: Kluwer), 643

CPSIA information can be obtained
at www.ICGtesting.com
Printed in the USA
BVHW022320130222
628173BV00027B/72

9 780750 322102